Dynamic Properties of Biomolecular Assemblies

Special Publication No. 74

Dynamic Properties of Biomolecular Assemblies

The Proceedings of a Symposium organised jointly by the Colloid and Interface Science Group of the Royal Society of Chemistry and the Techniques Group of the Biochemical Society

Nottingham, July 1988

Edited by

S. E. Harding
University of Nottingham

A. J. Rowe
University of Leicester

ROYAL
SOCIETY OF
CHEMISTRY

British Library Cataloguing in Publication Data

Dynamic properties of biomolecular assemblies: the
  proceedings of a symposium organised jointly by the
  Colloid and Interface Science Group of the Royal Society
  of Chemistry and the Techniques Group of the Biochemical
  Society, Nottingham, July 1988.
1. Polymers. Dynamics
I. Harding, S. E.   II. Rowe, A. J.   III. Colloid and
Interface Science Group   IV. Biochemical Society,
Techniques Group   V. Series
547.7′045413

ISBN 0-85186-896-7

© The Royal Society of Chemistry 1989

Published by The Royal Society of Chemistry,
Thomas Graham House, Cambridge, CB4 4WF

Printed in Great Britain by
St Edmundsbury Press Ltd, Bury St Edmunds, Suffolk

SD 5/11/90

Preface

The study of the behaviour of solutions and dispersions of large, biomolecular assemblies poses considerable intellectual and technological problems. Earlier work in this area had perforce to be confined to the explanation of intensive and rheological properties in terms of highly simplified models, and hence only a limited number of questions could meaningfully be asked.

In contrast, the last two decades have seen a bringing to fruition of a wide range of new methodologies, together with a growth in available computational power essential alike for adequate exploitation of older techniques. In addition, particularly but not solely in the field of protein assemblies, the detailed chemical and structural properties of large assemblies can now be relatively well described. Interactions and conformation changes thus become accessible in terms more amenable to empirical verification than heretofore.

The time thus seemed ripe to try to bring together workers from a range of disciplines who are currently involved in developing an area of growing importance. The organisers of the Joint Symposium of the UK Biochemical Society Techniques Group and Colloids and Interface Science Group of the Royal Society of Chemistry, held at Nottingham in July 1988 and upon which this book is based, were concerned to bring together scientists from both the 'colloid' and 'biophysical chemistry' traditions: who all too often in the past have worked in isolation from each others skills and technology. Note should also be taken of the growing interest in the area of industrial and applied science in the study of polymeric dispersions.

In this book we have sought to cover recent advances in a range of techniques that can be regarded as "Dynamic" (Hydrodynamics, NMR, Dynamic Light scattering, Fluorescence anisotropy decay and Electro-Optics), and their application to specific assemblies of biological macromolecules: assemblies involving proteins, nucleic acids and glycoconjugates. Some important advances in the capture and analysis of data (in many cases of the notoriously difficult 'multi-exponential' type) are identified. In vivo, assemblies of biological macromolecules do not always exist in a dilute solution environment - normally far from it - and some attention is paid to the dynamic properties of more concentrated dispersions.

The range of topics covered is clearly broad: readers make their own connections perhaps in some cases, but in the knowledge that the scientists who participated in the Symposium found multiple lively growths of common interest arising, and this in an area which is the essence of Biology: Dynamics. It is hoped that the reader will be similarly stimulated.

We are extremely grateful to all who helped in the organisation of the Joint Symposium, particularly to R. Dale and Doris Heriot of the Biochemical Society and to Dr. M. Hey of the Colloid & Interface Science Group of the Royal Society of Chemistry. We are also very grateful to the following organisations who provided generous financial support: Ciba-Geigy Pharmaceuticals, Malvern Instruments, Hoffmann-La Roche, Unilever Research, ICI Corporate Colloid Science Group, Polymer Laboratories, Hoescht, Pedigree (Mars) UK, Squibbs Technical Operations and Fisons Pharmaceuticals.

Stephen Harding

Arthur Rowe

Nottingham & Leicester

May, 1989

Contents

Part III: Dynamics of Glycoconjugates and Membranes

Part I: Techniques

1
Hydrodynamic Properties of Macromolecular Assemblies

By J. García de la Torre

DEPARTAMENTO DE QUÍMICA FÍSICA, UNIVERSIDAD DE MURCIA, E-30100 ESPINARDO, MURCIA, SPAIN.

1. AN OVERVIEW

This chapter is divided into two parts. The present one contains an overview of the foundations of the theory, the practical aspects of computational procedures and various applications. An abstract of the theory is presented in the second part, where all the equations needed for the calculations are given.

Introduction.

Since the development of the analytical ultracentrifuge and the viscosimetric techniques during the first half of this century, the measurements of hydrodynamic properties like the sedimentation coefficient, s, and the intrinsic viscosity, $[\eta]$, are among the simplest and most common ways to characterize the conformation and the dynamics (1,2) of macromolecular assemblies of biological origin, synthetic polymers and colloidal aggregates. Instrumental advances that took place in recent decades have made it possible to obtain other hydrodynamic information from the dynamic aspects of light scattering (3) and electro-optic (4) and spectroscopic (5) properties. Thus , the translational diffusion coefficient, D_t, (related to s through the well-know Svedberg formula (1,2)) is routinely determined from dynamic light scattering, and rotational diffusion coefficients (or relaxation times) can be derived from the time decay of properties like fluorescence anisotropy and electric birefringence or dichroism.

Macromolecular assemblies can be roughly classified into two broad classes: rigid and flexible. In the case of rigid macromolecules the hydrodynamic properties

3

depend on their size and shape only. Some of the
properties are exceedingly sensitive to these aspects
and are therefore a valuable source of structural
information. On the other hand, in the case of flexible
macromolecules the internal modes play an essential role
in the observable dynamics. This chapter is primarily
devoted to rigid macromolecules, although we shall also
indicate how rigid-body hydrodynamics can also be used
for flexible bodies.

A theoretical formalism is obviously needed in
order to extract structural information from
experimental values of hydrodynamic properties. Until
1967, the possible complexity of assemblies was largely
overlooked. In the analysis of properties very simple
models like axially symmetric ellipsoids were employed
(6). Such models are actually adequate for particles
like most globular proteins. The extension of the theory
to triaxial ellipsoids (7) has allowed a much more
elaborate description of the shape of such particles.

There are in any case some biological
macromolecules whose shape cannot be properly described
by biaxial or triaxial ellipsoids. Some typical examples
taken from the literature (8-12) are displayed in Fig.1.
In a pioneering work, Bloomfield et al. (13) proposed to
model such complex biomolecules using spherical
elements, or beads, as building blocks. The hydrodynamic
behavior of isolated spheres is well known, and the
hydrodynamic interaction between them was characterized
in the early works in an approximate way developed by
Kirkwood and others, and employed previously with some
success for flexible and rigid chains of beads (see
Yamakawa's monograph (14) for a review of Kirkwood theory
(15)). Thus, Bloomfield et al. used the approximate
double-sum formula of Kirkwood for D_t or s (eq. 22 and
23) and a similar equation (16) for rotational diffusion
(a simplified version of eq. 26-28).

Rigorous theory for rigid models.

A rigorous treatment of hydrodynamic properties of
bead models requires the evaluation of the three
components of the frictional forces at the beads as the
solutions of a system of 3N linear equations, where N is
the number of beads, in which the coefficients account
for the hydrodynamic interaction. Advances in the
description of hydrodynamic interaction between spheres
via Rotne-Prage-Yamakawa tensors (17,18) and increases
in availability of computational power prompted ten
years after Bloomfield's work the numerical treatment of

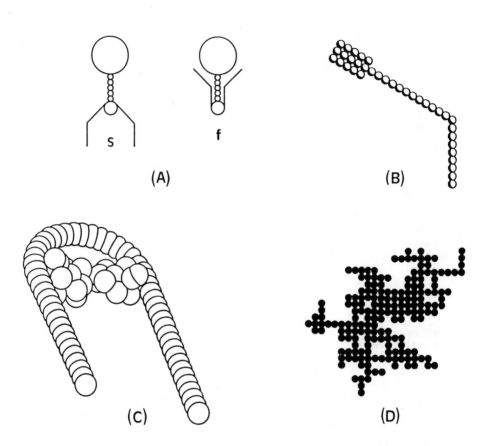

Figure 1. Examples of bead models for complex assemblies.
(A) T-even bacteriophage in slow (s) and fast (f) forms (ref. 8).
    Only two of the tailfibers (each of which is modelled with 64
    small beads) are shown.
(B) One-sixth of subcomponent C1q of human complement, according
    to Perkins. Other beautiful, related models are displayed in
    ref. 9
(C) DNA complex with cAMP promotor, according to Antonsiewicz and
    Porschke (10).
(D) Scheme of colloidal, fractal aggregate whose translational
    diffusivity has been studied by Meakin, Deutch and others (11,12)

bead models in a rigorous manner (19-22). Other aspects
including translation-rotation coupling and hydrodynamic
centres were refined in subsequent publications (23-25).
The state of this field was exhaustively reviewed by
Bloomfield and this author in 1981 (26). Other reviews
of a more limited coverage are available (27-29).

The fact noted in earliest papers (21-30) that the
original formalism gave $D_r=0$ and $[\eta]=0$ for a single
sphere was initially solved by replacing each bead by a
cubic array of eight smaller spheres (30,31). However,
this increases the computation time of the rigorous
treatment by a factor of about $8^3=512$. It was later
noticed by García de la Torre and Rodes (32) that a
simple correction that is proportional to the volume of
the particle (eq. 17) added to the diagonal components
of the rotational friction tensor accounts well for the
finite size of the elements. There is a similar volume
correction for $[\eta]$ (eq. 21,29,30) that is implicit in
papers of other workers (33-35). This correction, that
will be discussed in detail elsewhere (García de la
Torre et al., in preparation) is included in the bead
model treatment here for the first time. The volume
correction is essential in models in which one or a few
beads have a large fraction of the particle size. On the
other hand, when the model has many beads of similar
sizes (chain-like structures) the corrections are
unimportant.

In Part II of this paper we summarize the most
up-to-date version of the hydrodynamic theory of rigid
bead models, including volume corrections. The most
time-consuming step in the calculation of hydrodynamic
properties is the inversion of a 3Nx3N "supermatrix"
(eq.14), which requires a computation time of the order
of $\underline{a}N^3$, where $\underline{a}$ is a numerical constant. We have
recently noticed that in the formalism of our previous
papers, including our review articles (26,27), the
supermatrix $\underline{Q}$ to be inverted (eq.8 in ref.27) was not
symmetric. The formalism has been here changed, without
modifying its basis and its results, so that the
supermatrix which is inverted, $\underline{\mathcal{B}}$, is symmetric. This
reduces the time constant $\underline{a}$ by a factor of roughly two.
Goldstein (36) has recently proposed other computational
strategy for translational and rotational properties, in
which the system of linear equations which gives the
forces is solved avoiding matrix inversion, by means of
a Choleski decomposition. With this procedure, the time
constant $\underline{a}$ is reduced by a factor of about six. However,
it is not clear how the strategy of Goldstein can be
extended to the calculation of the intrinsic viscosity.

Approximate methods.

A detailed modeling of the shape of biological macromolecules requires usually a high number, N, of beads. As pointed out, however, the computer time of the rigorous computational method grows as N^3, and this places a practical limit in the level of modelling detail. An alternative is the use of approximate hydrodynamics. Concretely one can use double-sum formulas in which the properties are obtained directly as sums of pairwise contributions. A well-known example is the Kirkwood formula for D_t (eq. 22), which was improved by Bloomfield et al. for models with unequal beads (13) (eq. 23). We have recently derived a double-sum procedure (37) for the rotational friction tensor that leads directly to the rotational coefficients and relaxation times (eq. 24-28). A double-sum formula for $[\eta]$ derived by Tsuda (38), and a simplification of the latter derived by Freire and García de la Torre (39,40) are formulated in eqs. 29 and 30. We have here included for the first time the volume correction for $[\eta]$ in this formulas.

In all the double-sum methods, the computer time is proportional to N^2 (instead of N^3). Their inconvenience is that they are known to introduce an error that in some cases can be estimated. We have studied the errors of the approximate methods for several rigid structures like rods, rings and polygonal and polyhedral structures. In many cases of interest the errors are of a few percent only. Examples will be presented later.

Computer programs and examples of results

We have written two FORTRAN subprograms that implement the rigorous and the approximate methods, respectively. In both cases the input data are the number of beads, their Cartesian coordinates and their radii, and some input control parameters. The first lines of the two subprograms are the presented in Figures 2 and 3. The output results are organized in a few COMMON blocks as explained in the comments of the programs. The following units are used:

* u for lengths and dimensions (used in input data)
* $kT/6\pi\eta_0 u$ for translational diffusion
* $kT/6\pi\eta_0 u^3$ for rotational diffusion (its reciprocal for relaxation times).
* $N_A u^3/M$ for viscosity

here k is the Boltzmann constant, T is the absolute

```
                  SUBROUTINE TRV(IND,IP,B,NDIM)
C
C             SUBROUTINE TRV. VERSION 5
C      TRANSPORT PROPERTIES (TRANSLATION,ROTATION AND INTRINSIC
C      VISCOSITY) OF MODELS COMPOSED OF SPHERICAL ELEMENTS
C
C      WRITTEN BY J. GARCIA DE LA TORRE
C      DEPARTAMENTO DE QUIMICA FISICA, UNIVERSIDAD DE MURCIA
C      MURCIA ,SPAIN
C
C------DESCRIPTION OF INPUT PARAMETERS-------------------------
C
C      B IS THE 3NX3N SUPERMATRIX WHICH IS DIMENSIONED IN THE
C      CALLING PROGRAM AS B(x,x). NDIM IS THE VALUE SET FOR x
C      IN THE CALLING PROGRAM
C
C      IND=0  USE THE ORIGINAL OSEEN TENSOR
C --->  IND=1  USE THE MODIFIED OSEEN TENSOR (RECOMMENDED)
C      IND=9  NO HYDRODYNAMIC INTERACTIONS (ONLY FOR THEORETICAL
C             PURPOSES)
C
C
C      IP=0       NO PRINTING FROM THIS SUBROUTINE
C      IP=1       PRINTOUT OF FINAL RESULTS
C
C      THE CARTESIAN COORDINATES AND FRICTIONAL RADII OF BEADS ARE
C      ENTERED VIA A BLANK COMMON
C
C      THE FOLLOWING COMMON BLOCKS CONTAIN THE OUTPUT RESULTS:
C      TENS  ROT   TRAVIS  OTHER
C
C      THE OUTPUT VARIABLES ARE
C        DIFCD   TRANSLATIONAL DIFFUSION TENSOR AT CD
C        DIR     ROTATIONAL DIFFUSION TENSOR
C        CDICD   DIFFUSION COUPLING TENSOR AT CD
C        V       EIGENVALUES OF DIR
C        D       ONE THIRD OF THE TRACE OF DIR
C        DEL     ANISOTROPY OF DIR
C        REL     ROTATIONAL RELAXATION CONSTANSTS
C        TIM     ROTATIONAL RELAXATION TIMES
C        DT      TRANSLATIONAL DIFFUSION COEFFICIENT
C        FT      TRANSLATIONAL FRICTION COEFFICIENT
C        CD      CENTER OF DIFFUSION
C        VIS     INTRINSIC VISCOSITY
C        BETA    SCHERAGA-MANDELKERN PARAMETER
C        XVIS    VISCOSITY CENTER
C        XMASS   CENTER OF MASS (UNIFORM DENSITY)
C        RADGIR  RADIUS OF GYRATION
C        VOL     VOLUME
C
C      UNITS: U FOR DIMENSIONS (IN INPUT DATA)
C             K*T/6*PI*ETA*U FOR TRANSLATIONAL DIFFUSION
C             K*T/6*PI*ETA*U**3 FOR ROTATIONAL DIFFUSION
C             NA*U**3/M FOR VISCOSITY
C-------------------------------------------------------------
C
       COMMON N,X(100,3),E(100),R(100)
       COMMON/TENS/DIFCD(3,3),DIR(3,3),CDICD(3,3)
       COMMON/ROT/V(3),D,DEL,REL(5),TIM(5)
       COMMON/TRAVIS/DT,FT,CD(3),VIS,BETA,XVIS(3)
       COMMON/OTHER/XMASS(3),RADGIR,VOL
C
       DIMENSION B(NDIM,NDIM)
```

Figure 2. Heading lines of the FORTRAN subroutine TRV for rigorous calculation of hydrodynamic properties

```
      SUBROUTINE APROX(IP)
C
C     VERSION 3 OF THE OLD SUBRUTINE DOBSUM.
C     WRITTEN BY J. GARCIA DE LA TORRE,
C     DEPARTAMENTO DE QUIMICA FISICA, UNIVERSIDAD DE MURCIA
C
C     DOUBLE-SUM, APPROXIMATE METHODS FOR TRANSPORT PROPERTIES
C     (TRANSLATION, ROTATION AND INTRINSIC VISCOSITY) OF MODELS
C     COMPOSED OF SPHERICAL ELEMENTS
C
C     FORMULAS  OF KIRKWOOD AND BLOOMFIELD FOR DT
C     KIRKWOOD-TYPE METHOD (GARCIA DE LA TORRE ET AL. 1987) FOR
C     ROTATION, GARCIA DE LA TORRE-FREIRE AND TSUDA
C     FORMULAS FOR  VISCOSITY
C
C     IP=0 NO PRINTING FROM THIS SUBROUTINE
C     IP=1 PRINTOUT OF FINAL RESULTS
C
C     VOL: VOLUME
C     XMASS: CENTER OF MASS
C     CD: APPROXIMATE HYDRODYNAMIC CENTER
C     RADGIR: RADIUS OF GYRATION (UNIFORM DENSITY)
C
C     DT: TRASLATIONAL DIFFUSION COEFFICIENT (KIRKWOOD)
C     DTB:TRASLATIONAL DIFFUSION COEFFICIENT (BLOOMFIELD)
C
C     DIR: ROTATIONAL DIFFUSION TENSOR
C     DR1: EIGENVALUES OF THE ROTATIONAL DIFFUSION COEFFICIENT
C     D1:  MEAN ROTATIONAL DIFFUSION COEFFICIENT
C     DEL1: ANISOTROPY OF ROTATIONAL DIFFUSION
C
C     VISFG:  INTRINSIC VISCOSITY (FREIRE-GARCIA DE LA TORRE)
C     VISTSU: INTRINSIC VISCOSITY (TSUDA)
C
C     THIS PROGRAM USES THE SAME UNITS AND STRUCTURE OF COMMON BLOCKS AS
C     THE TRV PROGRAM. ARRAYS DIFCD AND CDICD ARE EMPTY AND THE
C     PLACE OF FT IN TRV IS NOW OCCUPIED BY DTB. VIS CONSTAINS VISTSU
C     AND VISFG IS IN XVIS(1). BETA IS CALCULATED FROM VISTSU
C
C
      COMMON N,X(100,3),E(100),R(100)
      COMMON/TENS/DIFCD(3,3),DIR(3,3),CDICD(3,3)
      COMMON/ROT/V(3),D,DEL,REL(5),TIM(5)
      COMMON/TRAVIS/DT,DTB,CD(3),VIS,BETA,XVIS(1)
      COMMON/OTHER/XMASS(3),RADGIR,VOL
```

Figure 3. Heading lines of FORTRAN subroutine APROX for the approximate double-sum formulas.

temperature and N_A is the Avogadro number. With given values of the solvent viscosity η_0 and the molecular weight, the above formulated factors are immediately evaluated. Multiplying the program results for the corresponding factors, one obtains the real value of the property.

As examples of the output of the computer programs, we reproduce those obtained for a bead model of the antibody molecule IgG3. Bead models of human antibodies have been used very recently by Burton, Davis and co-workers (41), who kindly provided us with the Cartesian coordinates. The models are displayed in Fig. 4, and the output of the computer program is presented in Fig. 5 and 6.

Various theoretical papers quoted up to this point contain results for the hydrodynamic properties of commonly found geometries. A typical case is that of oligomeric subunit structures having a polygonal or polyhedral shape. In Table I we report values of D_t, the reciprocal of the two relaxation times and $[\eta]$. The latter contains for the first time the volume correction. The results differ slightly from those reported in ref. 26 and 31 because volume corrections are here used instead of cubic substitution. We give only the longest and shortest relaxation times. For some of the structures there is an intermediate relaxation time that can be obtained from data in Ref. 40. Results for oligomeric structures in which the subunits are prolate ellipsoids have been reported (42).

In Table II we present the results for the six antibody molecules displayed in Fig.4. Tables I and II show that the approximate, double-sum methods perform rather well, with errors of a few percent only for a variety of structures, from the simplest dimer to a complex model like that of IgG3. Such good performance suggest that these methods could be the proper choice when a detailed modeling (and therefore a high N) is needed.

Viscoelasticity

Up to this point we have considered the zero-shear, zero-frequency intrinsic viscosity. A potential source of structural information is the frequency—dependent intrinsic viscosity $[\eta](\omega)$. Improving previous works of other authors (43,44), and noting that there is a weak coupling between the shear field and the translational and rotational motions of the body, Wegener (45) has

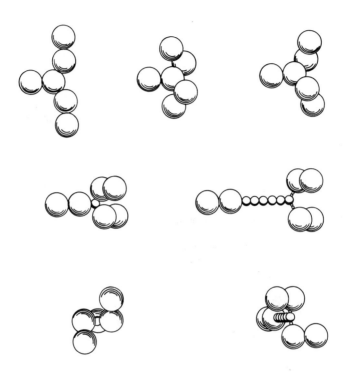

Figure 4. Bead models for human IgG antibody molecules, after Burton and coworkers (41)

```
SUBROUTINE TRV HAS ENTERED
VALUES OF THE CONTROL PARAMETERS
  IND    IP
   1     1

CARTESIAN COORDINATES AND RADII OF THE ELEMENTS
   1     -12.9000      0.0000       0.0000     2.0500
   2      -8.8000      0.0000       0.0000     2.0500
   3      -6.0000      0.0000       0.0000     0.7500
   4      -4.5000      0.0000       0.0000     0.7500
   5      -3.0000      0.0000       0.0000     0.7500
   6      -1.5000      0.0000       0.0000     0.7500
   7       0.0000      0.0000       0.0000     0.7500
   8       0.5000      0.0000       0.8660     0.2500
   9       0.7500      0.0000       1.2990     0.2500
  10       0.5000      0.0000      -0.8660     0.2500
  11       0.7500      0.0000      -1.2990     0.2500
  12       1.9000      0.0000      -3.2909     2.0500
  13       3.9500     -3.5507      -3.2909     2.0500
  14       1.9000      0.0000       3.2909     2.0500
  15       3.9500      3.5507       3.2909     2.0500

VOLUME=    225.62   RADIUS OF GYRATION=    7.4919

CENTER OF MASS (UNIFORM DENSITY)        -1.7162    0.0000     0.0000

CENTER OF DIFFUSION          -2.4717    0.0000     0.0000

TRANSLATIONAL DIFFUSION TENSOR
         0.189576     0.000000        0.000000
         0.000000     0.166808        0.005084
         0.000000     0.005084        0.169053

TRANSLATIONAL DIFFUSION COEFF.=    0.1751
TRANSLATIONAL FRICTION  COEFF.=    5.7095

ROTATIONAL DIFFUSION TENSOR
         0.004777     0.000000        0.000000
         0.000000     0.001757        0.000106
         0.000000     0.000106        0.001782

COUPLING TENSOR (DIFFUSION)
        -0.000425     0.000000        0.000000
         0.000000    -0.000121        0.000004
         0.000000     0.000004        0.000258

EIGENVALUES OF THE ROTATIONAL DIFFUSION TENSOR
   0.00166    0.00188    0.00478

TRACE OF ROT.DIF.TEN.=    0.00277
ANISOTROPY OF ROT.DIF.TEN.=    0.00301

RELAXATION CONSTANSTS
INDEX                  1          2          3          4          5
                    6D-2DEL    3D+3D1     3D+3D2     3D+3D3    6D+2DEL
ROTATIONAL CONST    0.01061    0.01331    0.01394    0.02265    0.02266
RELAXATION TIMES   94.27736   75.15629   71.71179   44.15849   44.13645

VISCOSITY CENTER       -2.2579    0.0000    0.0000

INTRINSIC VISCOSITY    2.577E+03

SCHERAGA-MANDELKERN PARAMETER         2.3177E+06
```

Figure 5. Output of the TRV subroutine for the model of the IgG3 antibody molecule displayed in Fig. 4

```
SUBROUTINE APROX

CARTESIAN COORDINATES AND RADII OF THE ELEMENTS
   1     -12.9000      0.0000      0.0000      2.0500
   2      -8.8000      0.0000      0.0000      2.0500
   3      -6.0000      0.0000      0.0000      0.7500
   4      -4.5000      0.0000      0.0000      0.7500
   5      -3.0000      0.0000      0.0000      0.7500
   6      -1.5000      0.0000      0.0000      0.7500
   7       0.0000      0.0000      0.0000      0.7500
   8       0.5000      0.0000      0.8660      0.2500
   9       0.7500      0.0000      1.2990      0.2500
  10       0.5000      0.0000     -0.8660      0.2500
  11       0.7500      0.0000     -1.2990      0.2500
  12       1.9000      0.0000     -3.2909      2.0500
  13       3.9500     -3.5507     -3.2909      2.0500
  14       1.9000      0.0000      3.2909      2.0500
  15       3.9500      3.5507      3.2909      2.0500

VOLUME=    225.62  RADIUS OF GYRATION=     7.4919

CENTER OF MASS (UNIFORM DENSITY)     -1.7162    0.0000    0.0000

APPROXIMATE HYDRODYNAMIC CENTER      -1.8255    0.0000    0.0000

TRANSLATIONAL DIFFUSION COEFFICIENTS
KIRKWOOD FORMULA:    0.2041
BLOOMFIELD FORMULA:    0.1816

ROTATIONAL DIFFUSION TENSOR
        0.004977      0.000000         0.000000
        0.000000      0.001927         0.000119
        0.000000      0.000119         0.001961

EIGENVALUES OF THE ROTATIONAL DIFFUSION TENSOR
   0.00182    0.00206    0.00498

TRACE OF ROT.DIF.TEN.=    0.00296
ANISOTROPY OF ROT.DIF.TEN.=    0.00304

RELAXATION CONSTANSTS
INDEX              1           2           3           4           5
                6D-2DEL      3D+3D1      3D+3D2      3D+3D3      6D+2DEL
ROTATIONAL CONST  0.01165    0.01434     0.01506     0.02380     0.02381
RELAXATION TIMES 85.84344   69.75770    66.40834    42.02211    41.99679

INTRINSIC VISCOSITY
TSUDA FORMULA:    2.2913E+03
FREIRE-GARCIA DE LA TORRE FORMULA:    2.2776E+03
SCHERAGA-MANDELKERN PARAMETER:    2.5968E+06(TSUDA)
```

Figure 6. The same as in Fig. 5 with the APROX subroutine.

Table I.- Hydrodynamic properties of oligomeric structures nomalized to the monomer values. Calculation were done with modified-Oseen tensors and volume correction. the percent error of the approximate methods of Kirkwood for D_t, García de la Torre *et al.* for $1/\tau$'s and Tsuda for $[\eta]$ are indicated.

Oligomer	D_t/D_{tsph}		$\tau_1^{-1}/\tau_{sph}^{-1}$		τ_5/τ_{sph}^{-1}		$[\eta]/[\eta]_{sph}$	
	EX	e%	EX	e%	EX	e%	EX	e%
monomer								
sphere	1.0	--	1.	--	1.	--	1.	--
dimer								
linear	0.75	0	0.21	-8	0.5	--	1.64	1
trimers								
triangle	0.66	-1	0.15	-6	0.18	-5	1.87	0
linear	0.60	-2	0.096	-1	0.33	--	2.10	0
tetramers								
square	0.58	-1	0.10	-1	0.12	-5	2.04	1
tetrahedron	0.61	-2	0.12	-4	0.14	4	1.97	3
linear	0.51	-3	0.051	-2	0.25	--	2.65	2
pentamers								
pentagon	0.52	-1	0.076	0	0.089	-4	2.20	2
bipiramid	0.57	-3					2.03	0
hexamers								
hexagon	0.47	0	0.057	-1	0.067	0	2.40	2
octahedron	0.54	-3	0.079	0	0.079	0	2.03	-3
trigonal prism	0.52	-3	0.074	-4	0.076	4	2.11	-1
linear	0.40	-3	0.020	-4	0.16	--	3.93	5
octamer								
cube	0.46	-4	0.045	-1	0.045	1	2.18	-1

Table II. Hydrodynamic properties and errors of approximate methods for the some antibody models of Burton and coworkers (Fig. 4). u=1 nm was the unit for length. The units for the properties are those formulated above.

Antibody	N	D_t			$1/\tau_1$		$[\eta]$		
		EX	%K	%B	EX	%G	EX	%T	%F
DOB	6	0.201	5	5	0.0164	3	1724	-5	-6
IgG1	7	0.205	9	6	0.0208	3	1610	-5	-7
IgG2	6	0.231	5	5	0.0265	2	1390	-5	-9
IgG3	15	0.204	16	3	0.0106	9	2577	-11	-12

EX: rigorous procedures. Approximate methods: K, Kirkwood; B, Bloomfield; G: García de la Torre *et al.*; T: Tsuda; F: Freire- García de la Torre.

presented a general treatment of the viscoelasticity of
rigid particles of arbitrary shape. He found that $[\eta](\omega)$
is the sum of five contributions having the usual form
in viscoelasticity, and the five time constants are just
those involved in the dynamics of electro-optic and
spectroscopic properties. Wegener has examined the real
and imaginary part of $[\eta](\omega)$ for a bent-rod model. The
frequency dependence is rather dependent on the
conformation. We are not aware, however, of experimental
studies analyzed in terms of his treatment.

For $\omega=0$, Wegener (45) obtained an expression for
$[\eta]$ that modifies slightly our formula, eq. 21. Other
authors had noticed that eq. 21 is not absolutely
rigorous since the mentioned coupling was not included
in its derivation (44,46). Indeed, it has been
demonstrated that eq. 21 is an upper bound for the exact
value of $[\eta]$. Iwata has found that eq. 21 is actually
exact for symmetric particles, and finally, Wegener
showed that in the worst cases, in which couplings are
expected to be strongest, eq. 21 deviate from the exact
result in less than 2%. Thus, eq.21 can be safely used
with the added advantage of its simplicity.

Higher-order hydrodynamic interaction

In the description of the hydrodynamic interaction
between a pair of isolated spheres, eq. 12 and 13 and
the treatment implicit in the volume correction for
rotation are all based as the truncation at the R_{ij}^{-3} of
series expansions of the $\underline{B}_{ij}$ matrices. When the
centre-to-centre distance is not appreciably larger than
the sizes of the subunits the truncation of the series
expansion may be premature. The following terms in the
series expansion of the hydrodynamic interaction between
two spheres are available (47,48). de Haen et al. (49)
have used pairwise interaction to derive new interaction
tensors and a modified-Kirkwood formula.

In an array of N spheres, hydrodynamic interaction
is actually a many-body problem and the pairwise
description is just an approximation. Series expansions
for B_{ii} and B_{ij} and related matrices that involve
three-sphere and even four-sphere terms have been
derived by Mazur and van Saarlos (50) and used in
bead-model calculations of translational coefficients by
Phillies (51). The conclusion reached by this author is
that additional terms in R_{ij}^{-n} have different signs, and
many-sphere contributions tend to cancel each other. In
short, Phillies' results for bead models with
sophisticated hydrodynamics interaction are rather close

to those obtained using just a Rotne-Prager-Yamakawa tensor, as we usually do, and he deduced that this simple treatment is also rather accurate. Rotational and viscoelastic properties, however, have not been explored in this regard, and as the problem of the hydrodynamic interaction between spheres is still alive (52), this topic could be reconsidered in the near future.

Brownian dynamics simulation

The whole body of bead-model hydrodynamic theory has a plausible alternative in the simulation of Brownian dynamics of the rigid assembly. If q_i and q_j are two of the six (translational and rotational) coordinates, according to Einstein relationship their covariance after a time step Δt is given by $\langle q_i q_j \rangle = 2D_{ij}\Delta t$. In single-step Brownian simulations (53,54) many Brownian steps are simulated and each of the six components of the diffusion matrix $\underline{\mathcal{D}}$ (eq. 2) can be so obtained. A simple simulation algorithm due to Ermack and McCammon (53,55,56) can be used, but a serious drawback is the need of accounting for the constraints that preserve the shape of the rigid model. Reinforcing those constraints without neglecting hydrodynamic interaction is a task that increases remarkably the computing time. I do not think that Brownian dynamics simulation is a competitive alternative to bead-model theory for rigid assemblies. However, there are related problems in which simulation is a very convenient (or even the only reliable) approach. One of them is the dynamics of segmentally flexible macromolecules, in which multistep simulation of Brownian trajectories gives after a time-correlation analysis, the relaxation P(t) of properties that for a rigid macromolecule would be given by eq. 7.

Shell-model calculations for continuous particles

An important application of bead-model calculation within the field of rigid-body hydrodynamics is the prediction of hydrodynamic properties of a continuous particle, which can be modelled as a shell of small beads that cover its whole surface. This idea was put forward by Bloomfield et al. when bead-model theory had not been fully developed as yet. More recently, the shell-model type of calculation has been applied to a variety of geometries including oligomers of spheres (57), ellipsoids (57) and cylinders (58). Owing to the ubiquity of the cylindrical shape among biopolymers, the systematic work on short cylinders (26,27,58,59) was particularly relevant since it showed and corrected the defect for moderate length-to-diameter ratio of the formulas presented in 1960 by Broersma (60). There are other, more recent applications of the shell-model idea. Roger and co-workers have calculated translational coefficients for disks and flat annular rings (61-62). Another shell-model calculation that may be of wide applicability is that of Allison *et al*. (63) for toroids, which is a shape frequently shown by condensed DNA. Wegener (64) has recently explored a calculational procedure developed by Youngreen and Acrivos (65) that resembles the shell-model concept, although with a more complex formalism. The applicability of this procedure has not been studied in depth.

Rigid-body treatment of flexible macromolecules

While the hydrodynamic theory of rigid particles or assemblies is fairly well established, as reviewed so far, the hydrodynamics of flexible particles is still problematic due to the complex interplay between statistical and dynamic aspects.

Following the development of rigid-body hydrodynamics in the late 1970's, Zimm (66) and our group (67) proposed a treatment for flexible systems in which the observable hydrodynamics properties are calculated averaging over the conformational space the values obtained for individual conformations that are regarded as instantaneously rigid. In most cases there are many degrees of conformational freedom and the averaging is carried out by Monte-Carlo simulation. It has been suggested that this treatment is only an approximation (68), but the results for flexible polymer chains are in excellent agreement with experiments.

In the field of biological macromolecules there are two cases in which this treatment is of interest. One of them is that of semiflexible, wormlike chains like DNA (70), in which a small flexibility takes place along the whole contour of the chain. Conceptually, this case is somewhat far from the idea of a rigid body and I do not describe it here, referring the reader to the work of Hagerman and Zimm (71). The other one is treated next.

Segmentally flexible macromolecules

The concept of segmental flexibility was put forward by Yguerabide (72) *et al*. to describe the flexibility of some macromolecules like immunoglobins, tRNA and myosin that are composed of a few, essentially rigid subunits connected by joints or hinges that are more or less flexible. Wegener (73,74), this author

(75-78) and their co-workers have presented an approach for completely flexible joints in which the low-Reynolds-number hydrodynamics (see part 2) is simply generalized from the six degrees of freedom for a particle with two subunits and an universal swivel (74,75). Bead-models expressions have been derived for the 9x9 tensors analogous to those in eq. 1-3. The problem of the proper hydrodynamic center of segmentally flexible macromolecules has also been solved (79).

A different approach is based on the rigid-body treatment. An early analysis of hydrodynamic properties of myosin and its parts used implicitly this treatment (80). Its utility to describe the rotational dynamics of segmentally flexible particle with partial flexibility was presented by Roitman and Zimm (81), who doubted the validity of Wegener's treatment.

In order to establish the validity or applicability of the two approaches, we undertook recently a Brownian dynamics simulation study (82) which avoids theory development and is based on first principles. The model studied, called "trumbbell" in the literature (81), had two arms and three beads (one of them is the joint). The bending potential was $V=Q\alpha^2/2$, where α is the angle between the two arms and Q is a flexibility parameter. We found that for Q=0, the rotational relaxation of the orientation of one arm, monitored for instance by the decay of anisotropy of fluorescence of a chromophore attached to the arm, is well described by Wegener's theory. On the other hand, for arbitrary Q, the collective rotation of the particle (concretely, of the end-to-end vector) monitored for instance by electric birefringence decay is well described by a rigid-body treatment for the longest relaxation time. Thus the two approaches are shown to be valid, although in different situations.

Perhaps the most extensively studied family of segmentally flexible molecules is that of myosin and its subfragments. The myosin molecule is depicted in Fig. 7. The simplest semiflexible subfragment is the myosin rod, or head less myosin, which has two rodlike subunits (S2 and LMM). The partial flexibility of the joint in the rod can be characterized by a bending energy that is quadratic in the bending angle. Table III summarizes the values of Q obtained from experimental data of various properties (of both equilibrium and transport) analyzed in terms of the rigid-body treatment. The corresponding values of the $\langle\alpha\rangle$ average are listed. The good agreement between the values obtained from different properties indicates the consistency and adequacy of the treatment. The values $Q \cong 0.6$ and $\langle \alpha \rangle \cong 60^{\circ}$ confirm the moderate flexibility of the rod which is supposed to play an essential role in the contraction of muscle(85).

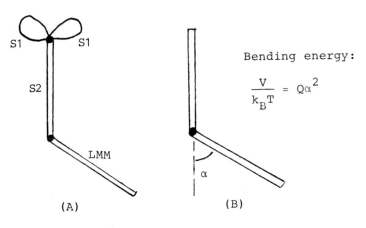

Bending energy:

$$\frac{V}{k_B T} = Q\alpha^2$$

(A) (B)

Figure 7. (A) Scheme of the whole myosin molecule, indicating the name of the subfragments. The black dots are the partially flexible joints.
(B) Myosin rod. The two arms, S2 and LMM are assumed to have the same length, 75 nm.

Table III. Analysis of the partial flexibility of myosin rod from experimental data interpreted in terms of the rigid-body treatment.

Property	Q	$\langle \alpha \rangle$	Ref.
Radius of gyration	0.42	63	84
Kerr constant	0.50	60	85
Intrinsic viscosity	0.70	54	83
Longest relaxation time in electric birefringence	0.65	55	83
Means:	0.57	58	

2. Summary of Theory.

Low Reynolds Number Hydrodynamics of Rigid Particles

A rigid particle has six degrees of freedom: three translations and three rotations, and its motion can be described by a six-dimensional velocity vector v, which contains three linear velocities and three angular velocities. If the fluid in which the particle is immersed is not at rest, the relevant velocities are the relative ones. The six-dimensional force vector, $\mathcal{F}_0$, that describe the frictional drag experienced by the particle contains three forces and three torques. Under low-Reynolds-number conditions, there is a linear relationship between $\mathcal{F}$ and v

$$\mathcal{F} = \Xi \cdot v \tag{1}$$

where Ξ is the 6x6 friction tensor.

The relationship between the deterministic, hydrodynamic domain, and the stochastic aspects of Brownian diffusion is established by a generalized form of the Einstein relationship. Let $\mathcal{D}$ denote a 6x6 diffusion matrix such that the covariance of the Brownian displacements of two of the six coordinates, q_i and q_j, per unit time is $2\mathcal{D}_{ij}$. The 6-dimensional Einstein equation is simply

$$\mathcal{D} = kT\Xi^{-1} \tag{2}$$

where k is Boltzman constant and T is the absolute temperature (86,87).

It is convenient to divide $\mathcal{D}$ and $\mathcal{R}$ into four 3x3 blocks, which correspond to translation (tt), rotation (rr) and coupling (tr). Thus, eq.2 is rewritten as

$$\begin{pmatrix} D_{tt} & D_{tr}^T \\ D_{tr} & D_{rr} \end{pmatrix} = kT \begin{pmatrix} \Xi_{tt} & \Xi_{tr}^T \\ \Xi_{tr} & \Xi_{rr} \end{pmatrix} \tag{3}$$

Superscript T denotes transposition. All the components of $\mathcal{D}$ and Ξ depend on the point at which they are referred or calculated except Ξ_{tt} and D_{rr}. The correct choice for such point is the centre of diffusion, whose position vector r_{OD} relative to the origin O at which the calculations are started is given by (24)

$$
\underline{r}_{OD} \equiv
\begin{pmatrix} x_{OD} \\ y_{OD} \\ z_{OD} \end{pmatrix}
=
\begin{pmatrix}
D_{rr}^{yy}+D_r^{zz} & -D_{rr}^{xy} & -D_{rr}^{xz} \\
-D_{rr}^{xy} & D_{rr}^{xx}+D_r^{zz} & -D_{rr}^{yz} \\
-D_{rr}^{xz} & -D_{rr}^{yz} & D_{rr}^{yy}+D_r^{zz}
\end{pmatrix}^{-1}
\begin{pmatrix}
D_{tr}^{yz}-D_{tr}^{zy} \\
D_{tr}^{zx}-D_{tr}^{xz} \\
D_{tr}^{xy}-D_{tr}^{yx}
\end{pmatrix}
\tag{4}
$$

The relationship that transforms the matrix $\underline{\underline{D}}_{tt,O}$ at O into its value at the centre of diffusion $\underline{\underline{D}}_{tt,D}$ is

$$
\underline{\underline{D}}_{tt,D} = \underline{\underline{D}}_{tt,O} - \underline{\underline{U}}_{OD} \cdot \underline{\underline{D}}_{rr} \cdot \underline{\underline{U}}_{OD} + \underline{\underline{D}}_{tr}^{T} \cdot \underline{\underline{U}}_{OD} - \underline{\underline{U}}_{OD} \cdot \underline{\underline{D}}_{tr}
\tag{5.a}
$$

where

$$
\underline{\underline{U}}_{OD} =
\begin{pmatrix}
0 & -z_{OD} & y_{OD} \\
z_{OD} & 0 & -x_{OD} \\
-y_{OD} & x_{OD} & 0
\end{pmatrix}
\tag{5.b}
$$

Observable quantities

The translational dynamics is simply characterized by the translational diffusion coefficient, D_t, which is just given by

$$
D_t = \frac{1}{3} \operatorname{Tr}(\underline{\underline{D}}_{D,tt})
\tag{6}
$$

The rotational dynamics is usually observed in the decay of electro-optic or spectroscopic properties. If $P(t)$ is the time course of one of, for instance, as the electric birefringence or the anisotropy of fluorescence, for a rigid particle this function is a sum of up to five exponentials

$$
P(t) = \sum_{i=1}^{5} A_i \exp(-t/\tau_i)
\tag{7}
$$

The reciprocals of the relaxation times, $1/\tau_i$ are given by the following formulas, which depend on the eigenvalues D_1, D_2 and D_3 of the origin-independent tensor $\underline{\underline{D}}_{rr}$ (89, 90):

$i=$	1	2	3	4	5
$1/\tau_i =$	$6D_r-2\Delta$	$3(D_r+D_1)$	$3(D_r+D_2)$	$3(D_r+D_3)$	$6(D_r+2\Delta)$

$$
\tag{8}
$$

where

$$Dr = \frac{1}{3} (D_1 + D_2 + D_3) \tag{9}$$

and

$$\Delta = (D_1^2 + D_2^2 + D_3^2 - D_1 D_2 - D_1 D_3 - D_2 D_3)^{1/2} \tag{10}$$

The A_i's are complicated function of the D_j's and the electro-optic or spectroscopic vectors and tensors expressed in the principal axes of rotational diffusion (90, 91). Another hydrodynamic property that is easily measurable is the intrinsic viscosity $[\eta]$ defined as the relative increase in the viscosity of the solvent due to the solute per unit of solute concentration for $[\eta]$ of a rigid body of arbitrary shape could be presented (92), but we prefer to present later a particularized form for bead models.

Hydrodynamics of bead models. Rigorous treatment

A rigid particle of arbitrary shape is modeled as an assembly of Stokes beads with frictional coefficients $\zeta_i = 6\pi\eta_0\sigma_i$ $(i=1,\ldots N)$ where η_0 is the viscosity of the solvent and σ_i is the radius of the bead. If the bead moves with velocity u_i in a fluid whose velocity at the centre of the bead would be v_i^o if the particle were absent, then, the frictional forces at the beads, F_i are related through a hydrodynamic-interaction equation

$$\sum_{j=1}^{N} \underline{B}_{ij} \cdot \underline{F}_j = \underline{u}_i - \underline{v}_i^o \tag{11}$$

where B_{ij} is a 3x3 matrix defined for i=j by $\underline{T}_{ii} = 1/\zeta_i$ and, for i≠j $\underline{B}_{ij} \equiv \underline{T}_{ij}$, where $\underline{T}_{ij}$ is the interaction tensor for which we usually adopt the Rotne-Prage-Yamakawa (17,18) expression as modified by Garcia de la Torre and Bloomfield for unequal beads (21):

$$\underline{T}_{ij} = \frac{1}{8\pi\eta_0 R_{ij}} \left[\underline{I} + \frac{R_{ij}R_{ij}}{R_{ij}} + \frac{(\sigma_i^2 + \sigma_j^2)}{R_{ij}} \left(\frac{1}{3}\underline{I} - \frac{R_{ij}R_{ij}}{R_{ij}} \right) \right] \tag{12}$$

Eq. 12 is valid only if $R_{ij} \geq \sigma_i + \sigma_j$. Otherwise, beads i and j overlap, and if both have the same radius σ, then $\underline{T}_{ij}$ is given by the Rotne-Prage formula (17),

$$\underset{=}{T}_{ij} = \frac{1}{6\pi\eta_0\sigma} \left[\left(1 - \frac{9}{32} \frac{R_{ij}}{\sigma}\right)\underset{=}{I} + \frac{3}{32} \frac{R_{ij}R_{ij}}{\sigma R_{ij}} \right] \tag{13}$$

In the earliest papers and the review article a matrix $\underset{=}{Q}_{ij} \equiv \zeta_i \underset{=}{B}_{ij}$ was used. We now prefer the present formulation because $\underset{=}{B}_{ij} = \underset{=}{B}_{ji}$. In the notation of ref.31, $\underset{=}{B}_{ij} = \underset{=}{B}_{ij}^{tt}$.

If $\underset{=}{\mathfrak{B}}$ is a "supermatrix" of dimension 3Nx3N, composed of NxN blocks $\underset{=}{B}_{ij}$ and its inverse $\underset{=}{\mathfrak{C}}$,

$$\underset{=}{\mathfrak{C}} = \underset{=}{\mathfrak{B}}^{-1} \tag{14}$$

is similarly partitioned into blocks $\underset{=}{C}_{ij}$, then we have for the blocks of the resistance matrix (eq.3) (78):

$$\underset{=}{\Xi}_{tt} = \sum_i \sum_j \underset{=}{C}_{ij} \tag{15}$$

$$\underset{=}{\Xi}_{tr} = \sum_i \sum_j \underset{=}{U}_i \cdot \underset{=}{C}_{ij} \tag{16}$$

$$\underset{=}{\Xi}_{rr} = 6\eta_0 V\underset{=}{I} - \sum_i \sum_j \underset{=}{U}_i \cdot \underset{=}{C}_{ij} \cdot \underset{=}{V}_j \tag{17}$$

In the oldest notation, $\underset{=}{C}_{ij}$ was $\zeta_j\underset{=}{S}_{ij}$. The advantage of the present formulation is that, unlike the supermatrices $\underset{=}{Q}$ and $\underset{=}{S}$, $\underset{=}{\mathfrak{C}}$ and $\underset{=}{\mathfrak{B}}$ are symmetric, and the inversion in eq.14 consumes less computer time. The $6\eta_0 V\underset{=}{I}$ term in the right-hand-side of eq.17 (in which V is the volume of the particle) was not included in the previous review (26), and is the volume-correction that accounts for the finite size of the frictional beads. In eqs. 15-17,

$$\underset{=}{U}_i = \begin{pmatrix} 0 & -z_i & y_i \\ z_i & 0 & -x_i \\ -y_i & x_i & 0 \end{pmatrix} \tag{18}$$

and

$$\underset{\sim}{r} \equiv (x_i, y_i, z_i) \tag{19}$$

are the coordinates of the centre of bead i in a system of coordinates centred at an arbitrary origin O. Using successively eqs. 3, 9, 10 and 8, the rotational relaxation times are calculated. Also, from eq. 15-18 and 4-6 the translational diffusion coefficient can be obtained.

The calculation of the intrinsic viscosity requires previously the location of the viscosity centre, whose position vector r_{0v} with respect to the origin O is given by the condition

$$\frac{\partial[\eta]}{\partial r_{0v}} = 0 \tag{20}$$

($[\eta]$ is a minimum at the viscosity centre).If r_i' is the position vector of the centre of bead i with respect to the viscosity centre, then $[\eta]$ is given by

$$[\eta] = \frac{5N_A V}{2M} + \frac{N_A}{M\eta_0} \sum_i \sum_j \left(\frac{1}{15} \sum_\alpha r_i'^\alpha c_{ij}^{\alpha\alpha} r_j'^\alpha \right.$$

$$+ \frac{1}{20} \sum_{\alpha \neq \beta} r_i'^\alpha c_{ij}^{\beta\alpha} r_j'^\beta - \frac{1}{30} \sum_{\alpha \neq \beta} r_i'^\alpha s_{ij}^{\alpha\beta} r_j'^\beta$$

$$\left. + \frac{1}{20} \sum_{\alpha \neq \beta} r_i'^\alpha c_{ij}^{\beta\beta} r_j'^\beta \right) \tag{21}$$

Eq. 20 actually leads to a system of 3 linear equations whose unknowns are the coordinates of the viscosity centre. This system is explicitly formulated in ref.22. The first term is the right-hand-side of eq.21 is the volume correction for the intrinsic viscosity (33,35) which is included here in the bead-model calculation of $[\eta]$ for the first time.

Approximate double-sum formulas

As commented previously, we restrict ourselves to approximate methods involving just a double-sum over pairwise contributions from the model's elements. Bloomfield et al. generalized the well-known Kirkwood formula for elements with unequal beads, finding

$$D_t = kT \left(\sum_i \zeta_i \right)^{-1} \left[1 + \left(6\pi\eta_0 \sum_i \zeta_i \right)^{-1} \sum_{i \neq j} \zeta_i \zeta_j R_{ij}^{-1} \right] \quad (22)$$

When there are appreciable difference between the sizes of the beads, an improved formula derived by Bloomfield et al. (13) is usually more accurate:

$$D_t = \frac{kT \left(\sum_i \sigma_i^3 + \sum_{i \neq j} \sigma_i^2 \sigma_j^2 R_{ij}^{-1} \right)}{6\pi\eta_0 \left(\sum_i \sigma_i^2 \right)^2} \quad (23)$$

Of the various approaches available for rotational quantities, the most complete and accurate is that recently presented by Garcia de la Torre et al. (37), in which $\underline{\underline{D}}_r$ is given by

$$\underline{\underline{D}}_r = kt\underline{\underline{\Xi}}_{c,rr}^{-1} \quad (24)$$

$\underline{\underline{\Xi}}_{c,rr}$ is the approximate rotational friction tensor referred to an approximate hydrodynamic centre, C defined by

$$\underset{\sim}{r}_{OC} = \sum_i \sigma_i \underset{\sim}{r}_i \Big/ \sum_i \sigma_i \quad (25)$$

Subtracting $\underset{\sim}{r}_{OC}$ from the position vectors $\underset{\sim}{r}_i$ we obtain the position vectors $\underset{\sim}{r}_i'$ referred to C. Then, we have

$$\underline{\underline{D}}_r = kT(\underline{\underline{A}}^{-1} + \underline{\underline{A}}^{-1} \cdot \underline{\underline{B}} \cdot \underline{\underline{A}}^{-1}) \quad (26)$$

where

$$\underline{\underline{A}} = 6V\eta_0\underline{\underline{I}} - \sum_i \zeta_i \underline{\underline{U}}_i' \cdot \underline{\underline{U}}_i' \quad (27)$$

and

$$\underline{\underline{B}} = -\sum_{i \neq j} \zeta_i \zeta_j \underline{\underline{U}}_i' \underline{\underline{T}}_{ij} \underline{\underline{U}}_j' \quad (28)$$

The $\underline{\underline{U}}_i$'s are constructed as usual from the coordinates of the vectors referred to C.

The intrinsic viscosity can be also computed from double-sum formulas. In the context of the approximation involved in these formulas, the viscosity centre is again C (eq. 25). Once the position vectors are referred to C, $[\eta]$ can be calculated either from the equation of Tsuda (38)

$$[\eta] = \frac{5N_AV}{2M} + \frac{N_A\pi}{M}\left(\sum \sigma_i r_i^2\right)\left\{1 + \frac{3}{4\left(\sum \sigma_i r_i^2\right)}\right.$$

$$\sum_{i\neq j}\sum \sigma_i\sigma_j\left(\frac{r_i r_j \cos\alpha_{ij}}{R_{ij}} + \frac{4r_i r_j(r_i^2 + r_j^2)\cos\alpha_{ij}}{10R_{ij}}\right.$$

$$\left.\left. - \frac{r_i^2 r_j^2(1+7\cos^2\alpha_{ij})}{10R_{ij}}\right)\right\} \tag{29}$$

where $\cos\alpha_{ij}=\underset{\sim}{r_i}'\cdot\underset{\sim}{r_j}'/r_i'r_j'$ is the cosine of the angle subtended by vectors $\underset{\sim}{r_i}'$ and $\underset{\sim}{r_j}'$. $[\eta]$ can also approximately calculated from a simpler equation derived by Freire and Garcia de la Torre (39,40), which may be nearly as precise as the previous one:

$$[\eta] = \frac{5N_AV}{2M} + \frac{N_A\pi}{M}\left(\sum_i \sigma_i r_i^2\right)^2\left(\sum_i \sigma_i r_i^2\right.$$

$$\left. + \sum_{i\neq j}\sum \sigma_i\sigma_j R_{ij}^{-1}\underset{\sim}{R_i}\underset{\sim}{R_j}\right)^{-1} \tag{30}$$

Viscoelasticity

The frequency dependent viscosity is formulated as

$$[\eta](\omega) = [\eta](\infty) + [\eta]_{Br}(\omega) \tag{31}$$

where $[\eta](\infty)$ is the nondiffusive contribution and $[\eta]_{Br}(\omega)$ is the diffusive, frequency-dependent contribution related to Brownian motion, which is given by

$$[\eta]_{Br}(\omega) = \sum_{j=1}^{5} \frac{E_j}{1+i\omega\tau_j} \tag{32}$$

The τ_j's are those presented in eq.8. The zero-frequency viscosity is

$$[\eta] = [\eta](\infty) + [\eta](0) \tag{33}$$

Expressions for the E_j's, $[\eta](\infty)$ and the resulting expression for $[\eta]$ (which improves very slightly eq. 21) are presented in Wegener's paper (43)

Cylinders

For cylinders of length L and diameter d, D_t and the rotational diffusion coefficients $D_r^\perp$ and $D_r^\parallel$ for rotation around transversal and longitudinal axes, respectively, can be formulated as

$$3\pi\eta_0 LD_t/kT = \ln(p) + \gamma \tag{34}$$

$$\pi\eta_0 L^3 D_r^\perp/kt = \ln(p) + \delta_\perp \tag{35}$$

$$A_0\pi\eta_0 L^3 D_r^\parallel/kT = p^2/(1+\delta_\parallel) \tag{36}$$

where $A_0=0.96$ (note that this datum was misprinted under eq.74 of ref.26). The Tirado-Garcia de la Torre results (58) for γ, $\delta_\perp$ and $\delta_\parallel$, fitted to interpolating polynomials by Garcia de la Torre and Bloomfield (26), read:

$$\gamma = 0.312 + 0.565/p + 0.100/p^2 \tag{37}$$

$$\delta_\perp = -0.662 + 0.917/p - 0.050/p^2 \tag{38}$$

$$\delta_\parallel = 0.677p - 0.183/p^2 \tag{39}$$

where $p=L/d$. Eq. 37-39 are particularly accurate for $p<20$. For cylinders there are three relaxation times whose reciprocals are $6D_r^\perp$, $5D_r^\perp+D_r^\parallel$ and $2D_r^\perp+4D_r^\parallel$. Usually only the first one is observed.

ACKNOWLEDGMENTS
      The author was introduced in this field by Prof.
Victor A. Bloomfield, whose encouragement and advice is
deeply appreciated. Works from our group reviewed in
this chapter were done thanks to the collaboration with
S.C. Harvey, J.J. Freire, M.C. López, M.M. Tirado, J.M.
García Bernal, A. Jiménez, P. Mellado, J.J. García
Molina, V. Rodes, F.G. Díaz and A. Iniesta, and were
supported by grants 4073-79, 555-81, 1409-82, and
PR0561-84 from the *Comisión Asesora de Investigación
Científica y Técnica.*

REFERENCES

1. C.R. Cantor and P.R. Schimmel, "Biophysical
    Chemistry", part II, Freeman, New York, 1979.
2. E.G. Richards, "An introduction to the Physical
    Properties of Large Molecules in Solution",
    Cambridge Univ. Press., Cambridge, 1980.
3. R. Pecora, Editor, "Dynamic Light Scattering",
    Plenum, New York, 1985.
4. S. Krause, Editor, "Molecular Electro-optics",
    Plenum, New York, 1981.
5. P.M. Bayley and R.E. Dale, Editors, "Spectroscopy
    and the Dynamics of Molecular Biological Systems",
    Academic, London, 1985.
6. H. Scheraga, "Protein Structure", Academic, New
    York, 1961.
7. S. Harding, chapter 2.
8. J. García de la Torre and V.A. Bloomfield,
    Biopolymers, 1977, 16, 1779.
9. S.J, Perkins, Biochem. J., 1985,228, 13
10. J. Antonsiewicz and D. Porschke, J. Biomol.
    Structure Dynamics, 1988, 5, 819.
11. Z.-Y. Chen, J.M. Deutch and P. Meakin, J. Chem.
    Phys., 1984, 80, 2982.
12. P. Meakin and J.M. Deutch, J. Chem. Phys., 1987, 86,
    4648.
13. V.A. Bloomfield, W.D. Dalton and K.E. Van Holde,
    Biopolymers, 1967, 5, 135.
14. H. Yamakawa "Modern Theory of Polymer Solutions",
    Harper & Row, New York, 1971
15. J.G. Kirkwood, J. Polym. Sci., 1954, 12, 1.
16. D.P. Filson and V.A. Bloomfield, Biochemistry, 1967,
    6, 1650.
17. J. Rotne and S. Prager, J. Chem. Phys., 1969, 50,
    4831.
18. H. Yamakawa, J. Chem. Phys., 1970, 53, 436.

19. J.A. McCammon and J.M. Deutch, Biopolymers, 1976, 15, 1397.
20. H. Nakajima and Y. Wada, Biopolymers, 1977, 16, 875.
21. J. García de la Torre and V.A. Bloomfield, Biopolymers, 1977, 16, 1747; 1765.
22. J. García de la Torre and V.A. Bloomfield, Biopolymers, 1978, 17, 1605.
23. J.M. García Bernal and J. García de la Torre, Biopolymers, 1980, 19, 751.
24. S.H. Harvey and J. García de la Torre, Macromolecules, 1980, 13, 960.
25. W.A. Wegener, Biopolymers, 1981, 20, 303.
26. J. García de la Torre and V.A. Bloomfield, Q. Rev. Biophys., 1981, 14, 81.
27. J. García de la Torre in Ref. 4, p. 81.
28. D.C. Teller, E. Swanson and C. de Haen, Meth. Enzymol., 1979, 61, 103.
29. V.A. Bloomfield, in Ref.5, p. 1.
30. R.W. Wilson and V.A. Bloomfield, Biopolymers, 1979, 18, 1205.
31. J.M. García Bernal and J. García de la Torre, Biopolymers, 1981, 20, 129.
32. J. García de la Torre and V. Rodes, J. Chem. Phys., 1983, 79, 2454.
33. U. Bianchi and A. Peterlin, J. Polym. Sci., 1968, 6, 1759.
34. S.I. Abdel-Khalik and R.B. Bird, Biopolymers, 1975, 14, 1915.
35. T. Yoshizaki, I. Nita and H. Yamakawa, Macromolecules, 1988, 21, 165.
36. R.F. Goldstein, J. Chem. Phys., 1985, 83, 2390.
37. J. García de la Torre, M.C. López Martínez and J.J. García Molina, Macromolecules, 1987, 20, 661.
38. K. Tsuda, Rheol. Acta, 1970, 9, 509; Polym. J., 1970, 1, 616.
39. J. Freire and J. García de la Torre, Macromolecules, 1983, 16, 331.
40. J. García de la Torre, M.C. López Martínez, M.M. Tirado and J.J. Freire, Macromolecules, 1983, 16, 112.
41. L. Gregory, K.G. Davis, B. Sheth, J. Boyd, R. Jefferis, C. Nave and D.R. Burton, Molecular Immunology, 1987, 24, 821.
42. M.C. López Martínez and J. García de la Torre, Biophys. Chem., 1983, 18, 268
43. H. Nakajima and Y. Wada, Biopolymers, 1978, 17, 2291.
44. K. Iwata, J. Chem. Phys., 1979, 71, 931; Biopolymers, 1980, 19, 125.
45. W.A. Wegener, Biopolymers, 1984, 23, 2243.
46. G. Wilemski and G. Tanaka, Macromolecules, 1981, 14,

1531.
47. P. Reuland, B.U. Felderhoff and R.B. Jones, Physica A, 1978, 365.
48. D.J. Jeffrey and Y. Onishi, J. Fluid Mech. 1984, 139, 261.
49. C. de Haen, R.A. Easterly and D.C. Teller, Biopolymers, 1983, 22, 1133.
50. P. Mazur and W. van Saarlos, Physica A, 1982, 115, 21.
51. G.D. Phillies, J. Chem. Phys., 1984, 81, 4046.
52. B.J. Yoon and S. Kim, J. Fluid. Mech. 1987, 185, 437.
53. S.A. Allison and J.A. McCammon, Biopolymers 1984, 23, 167.
54. E. Dickinson, S.A. Allison and J.A. McCammon, J. Chem. Soc. Faraday Trans II, 1985 81, 591.
55. D.L. Ermak and J.A. McCammon, J. Chem. Phys. 1978, 69, 1352.
56. E. Dickinson, Chem. Soc. Rev., 1985, 14, 421.
57. E. Swanson, D.C. Teller and C. de Haen, J. Chem. Phys., 1978, 68, 5097.
58. M.M. Tirado and J. García de la Torre, J. Chem. Phys., 1979, 71, 2581; J. Chem. Phys., 1980, 73, 1986.
59. J. García de la Torre, M.C. López Martínez and M.M. Tirado, Biopolymers, 1984, 23, 611; J. Chem. Phys., 1984, 81, 2047.
60. S. Broesma, J. Chem. Phys., 1960, 32, 1626; 1632; J. Chem. Phys., 1980, 53, 436.
61. R.P. Roger and R.G. Hussey, Phys. Fluids, 1982, 25, 915.
62. T.J. Ui, R.G. Hussey and R.P. Roger, Phys. Fluids, 1984, 27, 787.
63. S.A. Allison, R.A. Easterly and D.C. Teller, Biopolymers, 1980, 19, 1475.
64. W.A. Wegener, Biopolymers, 1986, 25, 627.
65. G.K. Youngreen and A. Acrivos, J. Fluid. Mech., 1985, 75, 377.
66. B.H. Zimm, Macromolecules 1982, 15, 520.
67. J. García de la Torre, A. Jimenez and J.J. Freire, Macromolecules, 1982, 15, 148.
68. M. Fixman, J. Chem. Phys., 1983, 78. 1594.
69. J. García de la Torre, M.C. López Martinez, M.M. Tirado and J.J. Freire, Macromolecules, 1984, 17, 2715.
70. V.A. Bloomfield, D.M. Crothers and I. Tinoco Jr., "Physical Chemistry of Nucleic Acids", chap. 5, Harper and Row, New York, 1974.
71. P.J. Hagerman and B.H. Zimm, Biopolymers, 1981, 20, 1481.
72. J. Yguerabide, H.F. Epstein and L. Stryer, J. Mol. Biol. 1970, 51, 573.

73. W.A. Wegener, R.M. Dowben and V.J. Koester, J. Chem. Phys., 1980, 73, 4086.
74. W.A. Wegener, Biopolymers, 1982, 21, 1409; J. Chem. Phys., 1982, 76, 6425.
75. S.C. Harvey, P. Mellado and J. García de la Torre, J. Chem. Phys., 1983, 78, 2081.
76. J.García de la Torre, P. Mellado and V. Rodes, Biopolymers, 1985, 24, 2145.
77. A. Iniesta and J. García de la Torre, Eur. Biophys. J., 1987, 14, 493.
78. P. Mellado, A. Iniesta, F.G. Diaz and J. García de la Torre, Biopolymers, 1988, in press.
79. W.A. Wegener, Macromolecules, 1985, 18, 2522.
80. J. García de la Torre and V.A. Bloomfield, Biochemistry, 1980, 19, 5118.
81. D.B. Roitman and B.H. Zimm, J. Chem. Phys., 1984, 81, 6333, 6348.
82. F.G. Diaz and J. García de la Torre, Biophysical J., 1988, in press.
83. A. Iniesta, F.G. Diaz and J. García de la Torre, Biophysical J., 1988, in press.
84. A. Solvez, A. Iniesta and J. García de la Torre, Int. J. Biol. Macromolecules, 1988, 10, 39.
85. A. Iniesta and J. García de la Torre, submitted.
86. S.C. Harvey and H. Cheung, in "Cell and Muscle Motility", vol. 2, R.M. Dowben and H. Cheung, Ed., p. 279, N.Y., 1982.
87. J. Happel and H. Brenner, "Low Reynolds Number Hydrodynamics", Nordhoff, Leiden, 1973.
88. H. Brenner, J. Colloid Interface Sci., 1967, 23, 407.
89. L.D. Favro, Phys. Rev., 1960, 119, 53.
90. G.G. Belford, R.L. Belford and G.Weber, Proc. Natl. Acad. Sci. U.S.A., 1972, 69, 1392.
91. W.A. Wegener, R.M. Dowben and V.J. Koester, J. Chem. Phys., 1979, 70, 622.
92. J.M. Batchelor, J. Fluid. Mech., 1978, 84, 237.

2

Modelling the Gross Conformation of Assemblies using Hydrodynamics: The Whole Body Approach

By S. E. Harding

DEPARTMENT OF APPLIED BIOCHEMISTRY AND FOOD SCIENCE, UNIVERSITY OF NOTTINGHAM, SUTTON BONINGTON, LE12 5RD, U.K.

1. INTRODUCTION

There are two basic approaches to modelling macromolecular conformation using hydrodynamic techniques. One, pioneered by Bloomfield, Garcia de la Torre and co-workers involves modelling the particles as arrays of spheres that interact in a way described by the Burgers-Oseen tensor: such advances have been described by Garcia de la Torre[1] earlier in this volume. In this Chapter I will describe the progress made over the last few years using the alternative 'whole body approach' in terms of general triaxial ellipsoids: viz. ellipsoids with three unequal axes. This extends the classical 'ellipsoid of revolution' approach of Perrin, Simha, Scheraga, Mandelkern and others to a model which allows a much greater variety of gross conformations. The experimental options available include various combinations of viscosity, sedimentation and rotational diffusional parameters together with measurements of radii of gyrations and molecular covolumes. I will discuss the problems of macromolecular solvation or "hydration" and the general approximation of a macromolecule to an ellipsoid.

2. WHY THE 'WHOLE BODY' APPROACH?

It is possible now to predict a number of hydrodynamic shape parameters for many complex structures - including flexible ones - by representing such structures as arrays of spheres that interact in a way described by the Burgers-Oseen (or modifications thereof) tensor. It is possible to predict from a model for such structures the sedimentation coefficient, intrinsic viscosity and rotational diffusional parameters, and by successively refining the model satisfactory agreement with experimental data can in general be achieved[1]. This type of modelling has had many interesting applications, and its major use has been in facilitating the choice between possible models based on prior information about the molecule (from e.g. x-ray crystallography). Specific examples have been given elsewhere in this volume by J. Garcia de la Torre[1], D. Porschke & J. Antosiewicz[2] (DNA-protein assemblies) and S. Perkins[3] (the immunological complement system).

With the undoubted power of the 'multiple sphere' or 'bead model' approach it could be questioned whether the alternative 'whole body' approach was now of value. By 'whole body' approach I mean starting off with just assuming a single general model - namely an ellipsoid - and then calculating the form of this 'equivalent' ellipsoid directly from one, two or three types of hydrodynamic measurement.

The usefulness of this latter 'whole body approach' lies in the following: Firstly there are two inherent limitations of the bead model approach. One is the 'uniqueness' problem where, although by successive refinement, an accurate fit to the observed experimental data can in general be obtained for a given hydrodynamic model, then depending on the complexity of the assumed model, there could be a large number of other models of comparable complexity which give an equally good fit to the data. The other limitation of the bead model approach is that important assumptions have to be made about macromolecular solvation or "hydration". This is normally an elusive parameter to measure (see, e.g. ref. 4) with the result that somewhat unsatisfactory - or at best very difficult - estimates have to be made: indeed, some hydrodynamic shape functions are a more sensitive function to "hydration" than to shape.

The second reason supporting the utility of the whole-body approach derives from the so-called 'biotech boom'. That is with the large amount of newly engineered macromolecules now being produced, a relatively quick estimate of their properties in solution compared to non-engineered macromolecules is highly desirable - and obviously this includes the ability to model the gross conformation of the macromolecule without any prior clues as to what this shape could be. Moreover the 'whole-body' approach - using general triaxial ellipsoids - can be employed without any need to 'assume' a hydration - by using hydration independent functions - other than it is assumed similar for two to three types of measurement.

I want to now consider the recent advances in the whole-body approach involving ellipsoidal shapes with 3 degrees of freedom - the general tri-axial ellipsoid - but before I do so it would be useful to briefly review the earlier 'ellipsoid of revolution' approach (Fig. 1).

Figure 1   The Ellipsoid of Revolution

Formed by rotating ellipse about either the major or the minor axis

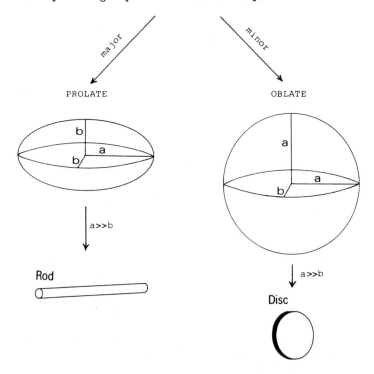

3. EARLY MODELS: THE SPHERE AND ELLIPSOID OF REVOLUTION

For over 80 years hydrodynamic (Greek: 'water moving') measurements have provided a valuable and relatively rapid way of estimating the dimensions of macromolecules - both synthetic and natural - in solution. The earlier calculations were based on spherical particles in terms of their frictional flow properties (sedimentation or diffusion) *through* a solution[5] and on their effect on the bulk viscous flow properties *of* the solution[6,7,8]. The advent of the analytical ultracentrifuge in the 1920's allowed the measurement of particle frictional ratios and the use of Perrin's[9] solutions in terms of ellipsoids, the latter being an extension of Stokes's solution[5] for spheres. Unfortunately, because of the complexity of the elliptic integral involved[9], macromolecules could only be modelled in terms of prolate or oblate 'ellipsoids of revolution'

Table 1   Hydrodynamic Shape Parameters for Ellipsoids of Revolution [semi-axes a,b,c and a>b=c (prolate) or a=b>c (oblate)]

Shape Parameter (as function of a/b)	Related Experimental Parameter	Ref
1. Bulk Solution Properties		
Viscosity increment, ν	Intrinsic viscosity, $[\eta]$	10,12,14
Reduced excluded volume, u_{red}	Thermodynamic 2nd virial coefficient, B (from light scattering, sedimentation equilibrium, etc.)	18,19
2. Translational Frictional Property		
Perrin function, P	Sedimentation coefficient, s; Translational diffusion coefficient, D	9,14
3. Rotational Frictional Property[a]		
'Reduced' birefringence decay constant, θ^{red}	[b]Electric birefringence decay constant, θ	21
Harmonic mean rotational relaxation time ratio, τ_h/τ_o	Harmonic mean rotational relaxation times (from steady state fluorescence depolarisation studies), τ_h	22
Fluorescence anisotropy depolarisation rotational relaxation time ratios, τ_i/τ_o (i = 1-3)	[c]Fluorescence anisotropy depolarisation relaxation times, τ_i	23

a. All use Perrin's[20] solutions for the rotational frictional ratios for ellipsoids.

b. There are two if optical axis does not coincide with geometric axis of ellipsoid of revolution. There are also two ($\theta_{\pm}$) for general ellipsoids (see below).

c. Three for ellipsoids of revolution, five for general ellipsoids.

<u>Table 2</u>   Hydration of Proteins Calculated from the Frictional Ratio, f/f_o. Axial ratios estimated from crystallographic dimensions of the proteins. From Squire & Himmel[15] & references cited therein.

Protein	Dimensions (Å)	a/b[a]	$P(a/b)$	f/f_o[b]	Hydration[c]
Basic trypsin inhibitor	29x19x19	1.53	1.016	1.447	0.86
Cytochrome C	25x25x37	1.48	1.014	1.116	0.24
Ribonuclease-A	38x28x22	1.52	1.016	1.290	0.73
Lysosyme	45x30x30	1.50	1.015	1.240	0.57
Myoglobin	44x44x25	1.76	1.028	1.170	0.35
Adenylate kinase	40x40x30	1.33	1.007	1.167	0.41
Trypsin	50x40x40	1.25	1.004	1.187	0.47
Bence-Jones protein REI	40x43x28	1.48	1.013	1.156	0.35
Chymotrypsinogen A	50x40x40	1.25	1.004	1.262	0.71
Elastase	55x40x38	1.41	1.010	1.214	0.53
Subtilisin	48x44x40	1.14	1.002	1.181	0.47
Carbonic anhydrase B	47x41x41	1.15	1.002	1.053	0.12
Superoxide dismutase	72x40x38	1.85	1.034	1.132	0.23
Carboxypeptidase A	50x42x38	1.25	1.004	1.063	0.14
Phosphoglycerate kinase	70x45x35	1.75	1.028	1.377	1.04
Concanavalin A	80x45x30	2.13	1.053	1.299	0.64
Hemoglobin, oxy	70x55x55	1.22	1.005	1.263	0.74
Bovine serum albumin	140x40x40	3.5	1.147	1.308	0.35
Malate dehydrogenase	64x64x45	1.42	1.011	1.344	1.00
Alcohol dehydrogenase	45x55x110	2.2	1.058	1.208	0.37
Lactate dehydrogenase	84x74x74	1.20	1.003	1.273	0.77

a. For the equivalent prolate or oblate ellipsoid. b. f_o in its use here corresponds to the frictional coefficient of an *anhydrous* spherical particle having the same mass and partial specific volume ($\bar{v}$) as the protein under consideration. In this definition $f/f_o = P.(f/f_h)$, where P is the Perrin function or 'frictional ratio due to shape'[15] and (f/f_h) a term due to hydration. c. Hydration, w (g solvent/g protein) $= [(f/f_h)^3]\bar{v}\rho_o$ where ρ_o is the solvent density.

and not general ellipsoids (3 unequal axes). Simha[10] extended Jeffrey's[11] earlier treatment for the viscous flow of solutions of ellipsoids of revolution to include the case of Brownian motion and gave an explicit relationship for the viscosity increment ν in terms of the semi-axes a, b of these ellipsoids. Saito[12] independently obtained the same result, suggesting that Simha had made an incorrect assumption (particles rotating with zero angular velocity) but had arrived at the correct result by making an 'error in calculation', a discrepancy resolved some 30 years later[13].

4. THE "HYDRATION" OR SOLVATION PROBLEM

An important problem in using the viscosity increment ν or the Perrin frictional ratio, P, is the experimental requirement of a value for the volume, V, of the macromolecule in solution, or equivalently protein "hydration", w^{14}. [Strictly speaking the term "solvation" should be used instead because other solvent species as well as water molecules can be trapped or bound to the macromolecule. However, since the term "hydration" has been used almost ubiquitously for several decades[14] we will hitherto follow the convention of using it to represent "associated solvent"]. Both functions are still commonly used direct, by using "assumed" hydration values. For example a hydration level, of 0.2-0.35 g water/g protein could be taken as typical for many globular proteins, although this value is still very arbitrary (see for example refs. 14, 15 for a discussion on this). Use of the other shape functions summarised in Table 1 also requires 'assumed hydrations'.

Unfortunately as we have mentioned above, hydration is a notoriously difficult parameter to measure with any meaningful precision. On the other hand if a reasonable estimate for the axial ratio of the molecule were known (from e.g. x-ray crystallography) then a hydration level could be calculated: Table 2 gives a summary of calculations performed by Squire & Himmel[15] to estimate protein hydration levels by using measured frictional ratios and estimated axial ratios from x-ray crystallography with the assumption that the protein has the same shape in solution as in crystallized form.

5. HYDRATION INDEPENDENT HYDRODYNAMIC SHAPE FUNCTIONS

The idea of combining analytically hydrodynamic shape functions dates back as long ago as 1953, with Scheraga & Mandelkern[16] who combined the relations of Perrin & Simha to give the well-used "β-function" (Table 2), extending the original 'graphical' approach of Oncley[17]. The β- function has proved however very insensitive to shape and has had most use as a quasi-constant shape parameter for determining molecular weights from intrinsic viscosity & sedimentation data.

Rotational diffusional phenomena provide in general parameters which

Table 3   'Compound' Hydration Independent Hydrodynamic Shape Parameters

Shape Parameter[a] (as function of a/b)	Comment	Ref
$\beta(\nu, P)$	Very poor sensitivity to axial ratio and high sensitivity to experimental error	16
$\Psi(\tau_h/\tau_o, P)$	Very poor sensitivity to axial ratio and high sensitivity to experimental error	34
$\psi(u_{red}, P)$	Very poor sensitivity to axial ratio and high sensitivity to experimental error	27
$R(\nu, P)$	Sensitive function at low axial ratio	30
$\Lambda(\tau_h/\tau_o, \nu)$	Very sensitive function, except at very low axial ratio (a/b $\leq$ 2.0)	35
$\Pi(u_{red}, \nu)$	Very sensitive function, except at very low axial ratio (a/b $\leq$ 2.0)	28

a: Source hydration dependent parameters are shown in parentheses.

are more sensitive functions of shape than the corresponding translational ones (Table 1): the hydration problem can also be accounted for by combination with appropriate translational parameters (either ν, or, P). This extra sensitivity comes however at a price: the two techniques commonly used, electric birefringence and fluorescence anisotropy depolarisation decay have some important practical limitations. In the case of electric birefringence, this is principally the requirement of having to use very low ionic strength solvents (due to conductivity problems)[24]; in the case of fluorescence depolarisation, the principal limitation is of internal rotation of chromophores or domains of the macromolecule relative to other parts[25]. Both techniques have difficulties of deconvolution of light source functions[24,26], and perhaps more seriously for asymmetric scatterers, both suffer from difficulties of resolution of multi-exponential decay terms and we will discuss this further below. A more recent development has been the use of a compound hydration-independent function (Λ) involving the *harmonic mean* rotational relaxation time[35] which largely avoids these problems and appears a sensitive function of axial ratio (Fig. 2).

Another recent development has been the use of compound shape functions involving molecular excluded volumes[27,28,29]. The molar covolume, U (ml.mol^{-1}) for a system of macromolecules can be obtained from the thermodynamic second virial coefficient, B (after correction for - or suppression of - charge effects). This covolume function is both a function of shape and hydration, but can be 'reduced' to give a function (u_{red}) of shape alone[28]. To experimentally determine it, a value for the hydration is still required, but again, the latter can be eliminated by combination with either the Perrin frictional ratio to give the hydration independent parameter ψ (ref. 27) or with the viscosity increment ν to give the hydration independent Π function[28]. Although the ψ function is very insensitive to shape - and rather disappointingly so, - the Π - function is on the other hand quite sensitive, and appears to be the most useful of the 'hydration independent' ellipsoid of revolution shape functions available.

Another 'hydration independent' parameter is the ratio R= $k_s/[\eta]$ where k_s is the sedimentation concentration dependence regression parameter and $[\eta]$ the intrinsic viscosity. It is known empirically[30,31,32] that R $\sim$ 1.6 for spheroidal particles (nb. after correction of sedimentation coefficients for solution density[31] - a higher value is obtained for coefficients corrected for solvent density) and < 1.6 for more asymmetric particles; after a number of assumptions and approximations a simple relation between R, ν and P has also been provided[33].

6. LIMITATIONS OF ELLIPSOIDS OF REVOLUTION. THE GENERAL ELLIPSOID

For many macromolecules the ellipsoid of revolution model can apparently give a reasonable representation of the gross conformation of macromolecules in solution. Indeed further examination of Table 2 will reveal that for many proteins, two of the three axial dimensions (derived from x-ray crystallographic data) are approximately equal. The disadvantages however of having to use a model with two axes equal are clear:

1. A decision has to be made *a priori* between the two types of ellipsoid of revolution (viz. prolate and oblate): virtually all of the usable hydration independent shape functions do not distinguish between the two (viz. they are not single-valued).

2. There are many classes of macromolecule which lie intermediate between a prolate shape (one long axis, two short) and an oblate (two long axes, one short).

As a result, hydrodynamicists have for a long time recognised the advantages of having a 'whole-body' model which does not have this restriction of two equal axes: the general triaxial ellipsoid (semi-axes a≥b≥c). This caters

Figure 2   Relative sensitivities of hydration independent shape functions
Broken line indicates minimum value this sensitivity must have if an axial ratio
precise to ± 20% is to be retrieved from the measured function, assumed precise
to ± 3%. ψ and Ψ not shown (very insensitive - close to baseline).

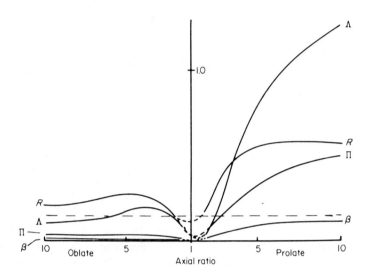

for a much wider range of shapes, from discs $(a{=}b{\gg}c)$, rods $(a \gg b{=}c)$, tapes $(a{\gg}b{\gg}c)$ and all intermediary shapes (Fig. 3).

The difficulty has been that the relation between the shape functions and the *two* axial ratios which characterise a general ellipsoid had either not been worked out (e.g. ν, u_{red}) or where computationally unavailable (satisfactory numerical routines for the evaluation of the elliptic integrals involved with many of the shape functions - notably the Perrin translational and rotational frictional ratio functions - and associated convergence problems). Over the last 15 years both of these problems have largely been addressed. Small & Isenberg[36] demonstrated that the Perrin elliptic integrals could be solved numerically using fast computers to evaluate the rotational and translational frictional ratio functions. The subsequent availability of the viscosity increment ν both numerically[37] and analytically[38] together with the reduced excluded volume[39] u_{red} for general 'tri-axial' ellipsoids, has now meant that a virtually complete set of hydration independent triaxial shape parameters are now available. A FORTRAN routine is available[40] for evaluating the set of hydrodynamic parameters for a particle for any given value of its axial

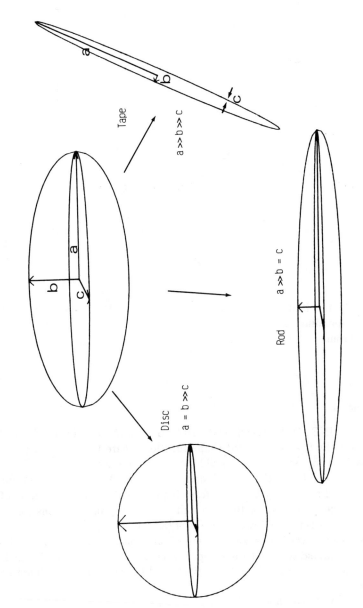

Figure 3   The General Triaxial Ellipsoid (semi-axes a≥b≥c) and its extremes

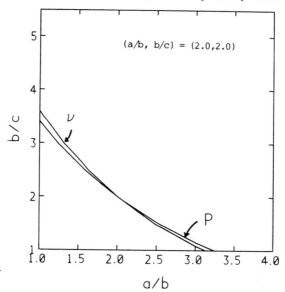

Figure 4   Plots of constant values for ν and P in the (a/b, b/c) plane corresponding to an (a/b, b/c) = (2.0, 2.0)
     From ref. 41

dimensions.

7. LINE SOLUTIONS: THE GRAPHICAL INTERSECTION METHOD

All the triaxial ellipsoid shape functions share the common property of having a line solution of possible values for the axial ratios (a/b, b/c) for any given value of the hydrodynamic function. A unique solution for these two axial ratios may be found from the intersection of two or more of these "line solutions" (Figs 4-10). Fig 4 illustrates two line solutions for ν and P for a hypothetical ellipsoid particle of (a/b, b/c) = (2.0, 2.0). Evidently this represents a very poor combination of functions, because of the shallowness of the intersection and their dependence on assumed values for hydration. To use the triaxial ellipsoid we have to find two suitable shape functions that are

1. hydration independent

Table 4   Hydration Independent Hydrodynamic Shape Parameters for Tri-axial Ellipsoids: Nature of Graphical Intersection

	Λ^a	G^b	$\delta_\pm^c$	R^d
Π	Poor inter-section	Good inter-section at high axial ratios	Good inter-section at high axial ratios	NE
Λ		Good inter-section at all axial ratios	NE	Good inter-section at all axial ratios
G			Good inter-section at low axial ratios	NE
$\delta_\pm$			.	Good inter-section at low ratios

a: assumes no internal rotation of chromophore or segmental rotation
b: from radius of gyration measurements (x-ray scattering or light scattering)
c: involves resolution of a two-term exponential decay
d: some approximations concerning concentration dependence of sedimentation coefficient
NE: not examined

2. experimentally measurable to a reasonable precision
3. are sensitive to shape (and insensitive to experimental error) and
4. give a reasonable intersection (i.e. as orthogonal as possible)
These criteria are quite restrictive, and in Table 4 we have summarised the intersection properties between the most useful functions. Formal definitions of these are given in the Appendix in the form of explicit relations in terms of the semi-axes a,b,c via the source parameters (ν, P, τ_h, $\theta_\pm^{red}$ and u_{red}), and also the corresponding experimental parameters.

Choice of shape function
In choosing two suitable hydration independent functions we try to avoid where possible those involving measurement of rotational diffusional or relaxational parameters. This is because almost always (except for those involving

the measurement of the harmonic mean relaxation time from steady state fluorescence depolarisation)[22], a step involving resolution of multi-exponential decay data is required. For example for general homogeneous ellipsoidal particles without an axis of symmetry there are two electric birefringence decay constants[42-44] and five fluorescence anisotropy depolarisation relaxation times[23,35]. Further, if solutions of macromolecules are not monodisperse, or if there is some self-association, there will be further exponential components. For example, two component electric birefringence relaxation data for haemocyanin solutions have been interpreted either as a polydisperse system of ellipsoids of revolution or as a monodisperse system of ellipsoids[43].

With electric birefringence there is the added complication of designing an instrument to have an adequate response time, deconvoluting the finite time to switch off the orienting field and problems of having to work at very low ionic strengths to avoid serious heating effects caused by the high electric fields used[24]. With fluorescence anisotropy depolarisation decay, besides having a formidable number of decay times to contend with (although in practice two-three are very similar) there is also a problem of deconvoluting the light source function from the decay data[26], together with the assumption of no significant internal rotation of the chromophore(s) or segmental rotation of parts of a given macromolecule relative to other regions of the same. A good demonstration of the segmental rotation problem has been given by Johnson & Mihalyi[25] for fibrinogen. The problems of multi-exponential resolution are not unique to macromolecular modelling and considerable attention has been paid and progress made as described elsewhere in this volume[47,48].

Despite the difficulties of rotational measurements an early attempt at modelling the triaxial conformation of (scallop) myosin light chains was made by Stafford & Szent-Gyorgi[49]: these workers made the approximation that of the five fluorescence anisotropy decay times, four similar 'fast' ones could be represented by a single harmonic mean, τ_h, which could be resolved from the 'slower' decay time τ_4. Although the ratio τ_h/τ_4 is hydration independent, at that time other hydration independent functions were not available, and so it was only possible to give limits for the axial ratios using a graphical combination of τ_h/τ_4 with the Perrin translational frictional function P, values for the latter evaluated using assumed values for the hydration. Although the intersection given for the case of 'no hydration' (Fig. 5) is rather meaningless, for more realistic values the intersection would appear to suggest an extended prolate shape of axial ratio between 5 and 6 to 1.

Use of hydration independent functions

Fortunately, shape function combinations are now available which largely avoid the difficulties referred to above, and a good example is the combination of the Π and G functions[50]. The availability of explicit relations between

Figure 5   Plots of constant values for τ_4/τ_h and P in the (a/b, b/c) plane for scallop myosin light chains.
P is given for 3 assumed values for the hydration, w(g H_2O/g protein). Redrawn from ref. 49.

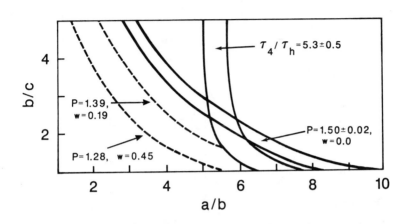

axial ratio (a/b, b/c) with the viscosity increment ν and reduced excluded volume, u_{red} for triaxial ellipsoids has enabled the Π function to be defined also[50]. As stated above Π can be obtained experimentally from measurements of intrinsic viscosity, and the thermodynamic 2nd virial coefficient (from e.g. light scattering, or sedimentation equilibrium - see, for example, the procedure described by Jeffrey *et al*[27]) after correction for Donnan effects.

The G-function has also been defined for triaxial ellipsoids[50] and can be obtained from the radius of gyration, again for example from light scattering or from low-angle x-ray scattering measurements. The G function is a very sensitive function to axial ratio, and an illustration of its use in conjunction with the Π function for a macromolecule of high axial ratio is given in Fig. 6 for myosin. In this particular example an attempt to model the gross conformation of the myosin molecule without prior assumptions about molecular hydration or using prior information from electron microscopy was made. Despite the extra degree of freedom the general ellipsoid gives, the myosin molecule appears as a prolate ellipsoid of (a/b, b/c) ≃ (80, 1). Nonewithstanding the difficulties of modelling an ellipsoid to a particle that in reality has a "lop- sided" end, this result is in good agreement with predicted results from electron microscopy, and would appear to suggest that local variations in particle shape (principally in the case of myosin the S1 heads) or flexibility

Figure 6   (a) Plots of constant values for Π and G in the (a/b, b/c) plane
for myosin. From ref. 50
(b) Gross conformation predicted

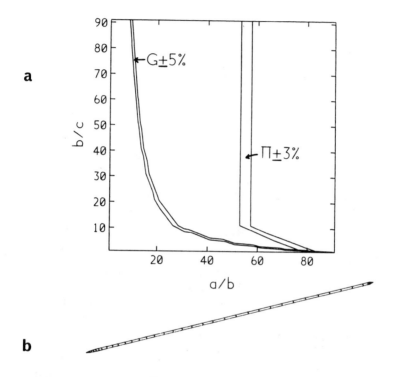

a

b

(the HMM/LMM interface) do not seriously distort estimates for the gross
conformation of the molecule using the triaxial ellipsoid in this way.

For the modelling of globular particles of low axial ratio (one axial ratio
$\lesssim$ 5) the intersection of G with Π is poor (largely through insensitivity of
the Π function in this region) and it is necessary to consider the use of other
combinations of hydration independent shape functions. A combination not
hitherto suggested, and which appears useful (Fig. 7) is the G - Λ combination. $Λ^{35,52}$ requires the measurement of intrinsic viscosity and the harmonic
mean rotation relaxation time, $τ_h$, a parameter which can be obtained from
steady state fluorescence depolarisation measurements without the need for
multi-exponential resolution. The G-Λ combination has, as yet not had any
practical application. The Λ function has however been combined[53] with the
R function, obtained from the ratio of the concentration dependence of the

Figure 7   Plots of constant values for G and Λ in the $(a/b, b/c)$ plane corresponding to an $(a/b, b/c) = (2.0, 2.0)$

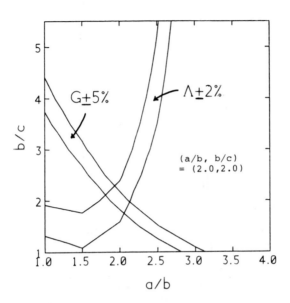

sedimentation coefficient, k_s, to the intrinsic viscosity, $[\eta]$. This provides a similarly sensitive intersection at low axial ratio and this combination of line solutions has been used to provide us with an indication of the likely mode of association of monomers of the neural protein neurophysin into dimers (Fig. 8)[53]. Again, as for myosin, despite the extra degree of freedom the general ellipsoid allows, the monomer still appears as a prolate model with two axes approximately equal $(a/b, b/c) \simeq (4.0, 1.0)$. For the dimer this reduces to an overall $(a/b, b/c)$ of $\simeq (2.8, 2.5)$, and the data therefore supports observations made earlier using ellipsoid of revolution models[54] that the association process is of a side-by-side rather than an end-to-end type.

In some applications the use of steady state fluorescence depolarisation functions can be limited, with the result that recourse to a function involving rotational diffusion is required. The $\delta_\pm$ are two such hydration independent functions obtained experimentally from the ratio of the (reduced) electric birefringence decay constants $\theta_\pm^{red}$ to the intrinsic viscosity, $[\eta]$. These functions also provide useful intersections with the G, Π and R functions [41,55] and Fig. 9 gives an example.

These 'useful intersections' however come at a price, namely the require-

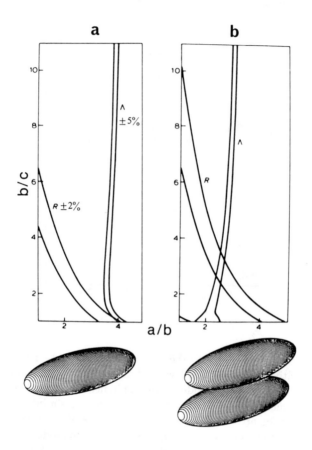

Figure 8   Plots of constant values for R and Λ in the (a/b, b/c) plane for
neurophysin monomers (a) and dimers (b). From ref. 53.

Figure 9   Plots of constant values of G, δ_+ and δ_- in the (a/b, b/c) plane corresponding to an (a/b, b/c) = (2.0, 2.0)

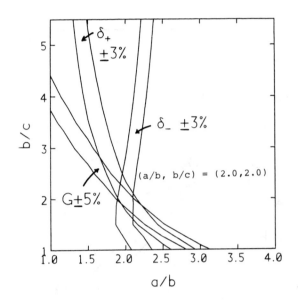

ment of extraction of the two exponential decay constants characterising general ellipsoids. We tried a whole series of procedures (non-linear least squares, Laplace transforms, method of moments etc.) on synthetic data with random error[52] but found the only reliable method for capturing the constants for data of "real" experimental precision were constrained least squares procedures, whereby estimates for the decay constants $\theta_\pm$ during an iteration process are constrained so that their corresponding values for $\delta_\pm$ in the (a/b, b/c) plane lie on a curve defined by another line solution (for example, G, Π or R). Fig. 10 gives an example of the "band" of allowed axial ratios (a/b, b/c) satisfactorily obtained in this way[41] for synthetic birefringence data with random expected 'experimental' noise. The limits of the band depend on the experimental precision of the constraining line solution and the birefringence data.

8. BEAD MODELS OR WHOLE-BODY MODELS?

Amongst the number of hydration independent shape functions now available, summarised in Table 4 and defined explicitly in the Appendix below it is hoped that there is at least one combination of functions suitable for mod-

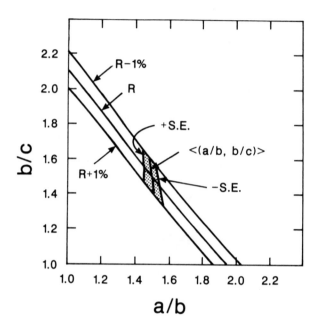

Figure 10   Constrained non-linear least squares fit of electric birefringence decay data. The plots are of constant R in the $(a/b, b/c)$ plane corresponding to an $(a/b, b/c)$ of $(1.5, 1.5)$. Shaded area corresponds to allowed band of axial ratios obtained by constraining the estimates for the δ_+ and δ_- functions to lie on the R curves. Simulated data, true $(a/b, b/c) = (1.5, 1.5)$; 0.1 deg. random standard error on the birefringence decay data.

elling the gross conformation of a given macromolecular system. It is also hoped that the 'bead model approach' for macromolecular modelling described by Garcia de la Torre elsewhere in this volume and the 'whole body' approach described here using triaxial ellipsoid shape functions prove complementary methods; the first, when close starting estimates for the structure are available from other sources and where the 'hydration' is known reasonably accurately, thereby facilitating a complex model; the second, when no prior shape information is available, and only the gross dimensions of the macromolecule in solution are required.

APPENDIX

Table 5 gives the formulation and relation to experimental parameters of the principal hydration dependent (i.e. requiring knowledge of the volume V of a particle, including associated solvent or 'hydration') and hydration independent shape functions for general triaxial ellipsoids of semi-axes (a $\geq$ b $\geq$ c). In these formulae:

1. In the equation for ν (the viscosity increment), δ is a small term ($\lesssim$ 1% of the other term) given by

$$\delta = -\frac{1}{5abc}\left[\frac{(\frac{a^2-b^2}{a^2\alpha_1+b^2\alpha_2} + \frac{b^2-c^2}{b^2\alpha_2+c^2\alpha_3} + \frac{c^2-a^2}{c^2\alpha_3+a^2\alpha_1})^2}{(\frac{a^2+b^2}{a^2\alpha_1+b^2\alpha_2} + \frac{b^2+c^2}{b^2\alpha_2+c^2\alpha_3} + \frac{c^2+a^2}{c^2\alpha_3+a^2\alpha_1})}\right]$$

2. In the equation for the reduced excluded volume, u_{red}, the R and S are double integrals given by

$$R = \frac{2}{3\pi}\int_0^{\pi/2}\int_0^{\pi/2} cos\ u\ du\ dv \left\{\left(\frac{a}{bc} + \frac{b}{ac} + \frac{c}{ab}\right)\Delta^2\right.$$

$$-sin^2\ v\ cos^2\ v\ cos^2 u\ \Delta^4\left(\frac{1}{a^2} - \frac{1}{b^2}\right)\frac{1}{c}\left(\frac{b}{a} - \frac{a}{b}\right)$$

$$-sin^2\ u\ cos^2\ u\ \Delta^4\left(\frac{cos^2\ v}{a^2} + \frac{sin^2\ v}{b^2} - \frac{1}{c^2}\right)$$

$$\left.\cdot\left[\frac{1}{c}\left(\frac{b\ cos^2\ v}{a} + \frac{a\ sin^2\ v}{b}\right) + \frac{c}{ab} - \frac{b}{ac} - \frac{a}{bc}\right]\right\}$$

Table 5   Shape functions for General Triaxial Ellipsoids

Function	Formulation	Relation to Experimental Parameters	Ref.
$*\nu$	$\frac{1}{abc}\left\{\frac{4(\alpha_7+\alpha_8+\alpha_9)}{15(\alpha_8\alpha_9+\alpha_9\alpha_7+\alpha_7\alpha_8)}\right.$ $+\frac{1}{5}\left[\frac{\alpha_2+\alpha_3}{\alpha_4(b^2\alpha_2+c^2\alpha_3)}\right.$ $+\frac{\alpha_3+\alpha_1}{\alpha_5(c^2\alpha_3+a^2\alpha_1)}$ $\left.\left.+\frac{\alpha_1+\alpha_2}{\alpha_6(a^2\alpha_1+b^2\alpha_2)}\right]\right\}+\delta,$	$[\eta].M_r/(V.N_A)$	38,42 46,52
$*P$	$2/\{(abc)^{\frac{1}{3}}/\alpha_{10}\}$	$[(\bar{v}M_r)/(VN_A)](f/f_o)$	9,35,52
$*u_{red}$	$2+(\frac{3}{2\pi abc})R.S,$	$\frac{U}{N_AV}\equiv\frac{1}{N_AV}.$ $[2BM_r^2-f(Z,I)]$	39
$*\tau_h/\tau_o$	$1/[a^2\alpha_1+b^2\alpha_2+c^2\alpha_3]$	$(kT\tau_h)/(3\eta_oV)$	52,53
$*\theta_\pm^{red}$	$\frac{abc}{12}\left\{(\frac{1}{Q_a}+\frac{1}{Q_b}+\frac{1}{Q_c})\right.$ $\pm\left[(\frac{1}{Q_a^2}+\frac{1}{Q_b^2}+\frac{1}{Q_c^2})\right.$ $-(\frac{1}{Q_aQ_b}+\frac{1}{Q_bQ_c}$ $\left.\left.+\frac{1}{Q_cQ_a})\right]^{\frac{1}{2}}\right\}$	$(\eta_o/kT).V.\theta_\pm$	43,44, 52,40
β	$(N_A^{1/3}/16200\pi^2)^{1/3}.$ $(\nu^{1/3}/P)$	$N_As[\eta]^{1/3}\eta_o/$ $[M_r^{2/3}(1-\bar{v}\rho_o)100^{\frac{1}{3}}]$	41,51,52
Π	u_{red}/ν	$U/([\eta].M_r)$ $\equiv[2BM/[\eta]]-$ $f(Z,I)/([\eta]M_r)$	50
G	$(1/5)[(a^2+b^2+c^2)$ $/(abc)^{2/3}]$	$[(4\pi N_A)/(3\bar{v}M_r)]^{2/3}.$ R_g^2	50
Λ	$\nu/(\tau_h/\tau_o)$	$(3\eta_o[\eta]M_r)/(N_AkT\tau_h)$	53
R	$2[1+P^3]/\nu$	$k_s/[\eta]$	41
$\delta_\pm$	$6\theta_\pm^{red}\nu$	$(6\eta_o/N_AkT).[\eta].M_r.\theta_\pm$	41

∗ : hydration dependent

and

$$S = \frac{8}{3} \int_0^{\pi/2} \int_0^{\pi/2} \cos u \; du \; dv \left\{ \left(\frac{bc}{a} + \frac{ca}{b} + \frac{ab}{c} \right) \Delta \right.$$

$$- \sin^2 v \cos^2 v \cos^2 u \; \Delta^3 \; c \left(\frac{b}{a} - \frac{a}{b} \right) \left(\frac{1}{a^2} - \frac{1}{b^2} \right)$$

$$- \sin^2 u \cos^2 u \; \Delta^3 \left(\frac{\cos^2 v}{a^2} + \frac{\sin^2 v}{b^2} - \frac{1}{c^2} \right)$$

$$\left. \cdot \left[c \left(\frac{b \cos^2 v}{a} + \frac{a \sin^2 v}{b} \right) - \frac{ab}{c} \right] \right\}$$

where

$$\Delta^{-2} = \frac{\cos^2 u \cos^2 v}{a^2} + \frac{\cos^2 u \sin^2 v}{b^2} + \frac{\sin^2 u}{c^2}$$

3. In the equation for the reduced decay constants $\theta_{\pm}^{red}$, the terms Q_a, Q_b and Q_c are given by

$$Q_a = \frac{b^2 + c^2}{b^2 \alpha_2 + c^2 \alpha_3}, Q_b = \frac{c^2 + a^2}{c^2 \alpha_3 + a^2 \alpha_1}, Q_c = \frac{a^2 + b^2}{a^2 \alpha_1 + b^2 \alpha_2}$$

4. The elliptic integrals $\alpha_1 - \alpha_{10}$ are given by

$$\alpha_1 = \int_0^\infty \frac{d\lambda}{(a^2 + \lambda)\Delta}; \alpha_2 = \int_0^\infty \frac{d\lambda}{(b^2 + \lambda)\Delta}; \alpha_3 = \int_0^\infty \frac{d\lambda}{(c^2 + \lambda)\Delta}$$

$$\alpha_4 = \int_0^\infty \frac{d\lambda}{(b^2 + \lambda)(c^2 + \lambda)\Delta}; \alpha_7 = \int_0^\infty \frac{\lambda d\lambda}{(b^2 + \lambda)(c^2 + \lambda)\Delta}$$

$$\alpha_5 = \int_0^\infty \frac{d\lambda}{(c^2 + \lambda)(a^2 + \lambda)\Delta}; \alpha_8 = \int_0^\infty \frac{\lambda d\lambda}{(c^2 + \lambda)(a^2 + \lambda)\Delta}$$

$$\alpha_6 = \int_0^\infty \frac{d\lambda}{(a^2 + \lambda)(b^2 + \lambda)\Delta}; \alpha_9 = \int_0^\infty \frac{\lambda d\lambda}{(a^2 + \lambda)(b^2 + \lambda)\Delta}$$

$$\alpha_{10} = \int_0^\infty \frac{d\lambda}{\Delta} \quad where \; \Delta = [(a^2 + \lambda)(b^2 + \lambda)(c^2 + \lambda)]^{1/2}$$

and λ is a dummy variable

These integrals can be solved numerically using standard computational packages without convergence problems - see *e.g.* ref 40 for a simple FORTRAN program illustrating their use.

5. The following experimental parameters are:

$[\eta]$ Intrinsic viscosity (ml/g)

M_r Molecular weight (g/mol)

V   Particle volume (including associated solvent or 'hydration') (ml)

N_A Avogadro's number (mol^{-1})

$\bar{v}$ Partial specific volume (ml/g)

(f/f_o) Frictional ratio. Following the most popular convention (ref 14), f_o refers to the frictional coefficient of a spherical particle of the same mass and *anhydrous* volume as the macromolecule whose frictional coefficient is f. This differs from our previous usage, which is that of Scheraga and Mandelkern[16] where f_o refers to a sphere of the same *hydrated* volume.

U   Molar covolume (ml/mol)

B   Thermodynamic second virial coefficient (ml.mol.g^{-2})

f(Z,I)   Function of macromolecular charge Z and solution ionic strength, I; f= 0 at the isoelectric pH for proteins, and → 0 as I is increased

k   Boltzmann constant (erg. K^{-1})

T   Absolute temperature (K)

τ_h Harmonic mean rotational relaxation time (sec)

η_o Solvent viscosity (Poise)

θ_+, θ_- Electric birefringence decay constants (2 for monodisperse solution of triaxial ellipsoids) (sec^{-1})

s   Sedimentation coefficient (sec)

R_g Radius of Gyration (cm)

k_s Concentration dependence sedimentation regression coefficient (ml/g)

REFERENCES

1. J.G. Garcia de la Torre, Chapter 1, this volume.
2. D. Porschke, Chapter 6, this volume.
3. S.J. Perkins, Chapter 15, this volume.
4. I.D. Kuntz and W. Kauzmann, *Adv. Prot. Chem*, 1974, *28*, 239
5. Sir G. Stokes, *Trans. Cambridge Phil. Soc.*, 1847, *8*, 287 and 1851, *9*, 8
6. A. Einstein, *Ann. Physik.*, 1906, *19*, 289
7. A. Einstein, *Ann. Physik.*, 1911, *34*, 591
8. A. Einstein, 'Investigations of the Theory of Brownian Movement' (Ed. R. Furth), Dover Publications, New York, 1956.
9. F. Perrin, *J. Phys. Radium*, 1936, *7*, 1
10. R. Simha, *J. Phys. Chem.*, 1940, *44*, 25
11. G.B. Jeffrey, *Proc. Roy. Soc. London Ser. A*, 1922, *102*, 161
12. N. Saito, *J. Phys. Soc. Japan*, 1951, *6*, 297
13. S.E. Harding, M. Dampier and A.J. Rowe, *Biophys. Chem.*, 1982, *15*, 205
14. C. Tanford, 'Physical Chemistry of Macromolecules', Chap. 6, J. Wiley and Sons, New York, 1961
15. P.G. Squire and M. Himmel, *Arch. Biochem. Biophys.*, 1979, *196*, 165
16. H.A. Scheraga and L. Mandelkern, *J. Am. Chem. Soc.*, 1953, *79*, 179
17. J.L. Oncley, *Ann. N.Y. Acad. Sci.*, 1941, *41*, 121
18. A.G. Ogston and D.J. Winzor, *J. Phys. Chem.*, 1975, *79*, 2496
19. L.W. Nichol, P.D. Jeffrey and D.J. Winzor, *J. Phys. Chem.*, 1976, *80*, 648
20. F. Perrin, *J. Phys. Radium*, 1934, *5*, 497
21. H. Benoit, *Ann. Phys.*, 1951, *6*, 561
22. G. Weber, *Adv. Prot. Chem.*, 1953, *8*, 415
23. C.R. Cantor and T. Tao, *Proc. Nucl. Acid. Res.*, 1971, *2*, 31
24. E. Fredericq and C. Houssier, 'Electric Dichroism and Electric Birefringence', Clarendon Press, Oxford, 1973
25. P. Johnson and E. Mihalyi, *Biochim. Biophys. Acta*, 1965, *102*, 476
26. I. Isenberg, R.D. Dyson and R. Hanson, *Biophys. J.*, 1973, *13*, 1090
27. P.D. Jeffrey, L.W. Nichol, D.R. Turner and D.J. Winzor, *J. Phys. Chem.*, 1977, *81*, 776
28. S.E. Harding, *Int. J. Biol. Macromol.*, 1981, *3*, 340
29. L.W. Nichol and D.J. Winzor, *Meth. Enzymol.*, 1985, *117*, 182
30. P.Y. Cheng and H.K. Schachman, *J. Polym. Sci.*, 1955, *16*, 1930
31. H.K. Schachman, 'Ultracentrifugation in Biochemistry', Chapter 4, Academic Press, New York, 1959
32. J.M. Creeth and C.G. Knight, *Biochim. Biophys. Acta*, 1965, *102*, 549
33. A.J. Rowe, *Biopolymers*, 1977, *16*, 2595
34. Squire, P.G., *Biochim. Biophys. Acta*, 1970, *221* 425; *Electro. Opt. Ser.*, 1978, *2*, 569

35. S.E. Harding, *Biochem. J.*, 1980, *189*, 359
36. E.W. Small and I. Isenberg, *Biopolymers*, 1977, *16* 1907
37. J.M. Rallison, *J. Fluid Mech.*, 1978, *84*, 237
38. S.E. Harding, M. Dampier and A.J. Rowe, *J. Coll. Int. Sci.*, 1981, *79*, 7; (see also *ibid*), *IRCS (Int. Res. Commun. Syst.) Med. Sci.*, 1979, *7*, 33
39. J.M. Rallison and S.E. Harding, *J. Coll. Int. Sci.*, 1985, *103*, 284
40. S.E. Harding, *Comput. Biol. Med.*, 1982, *12*, 75
41. S.E. Harding and A.J. Rowe, *Biopolymers*, 1983, *22*, 1813 (see also *ibid*, 1984, *23*, 843)
42. W.A. Wegener, *Biopolymers*, 1984, *23*, 2243
43. D. Ridgeway, *J. Am. Chem. Soc.*, 1968, *90*, 18
44. W.A. Wegener, R.M. Dowben and V.J. Koester, *J. Chem. Phys.*, 1979, *70*, 62?
45. S.E. Harding and A.J. Rowe, *Int. J. Biol. Macromol.*, 1982, *4*, 161
46. S.H. Haber and H. Brenner, *J. Coll. Int. Sci.*, *97*, 496
47. L. Brand, Chapter 7, this volume
48. A.K. Livesey, Chapter 8, this volume
49. W.F. Stafford, III and A.G. Szent-Gyorgi, *Biochemistry*, 1978, *17*, 620
50. S.E. Harding, *Biophys. J.*, 1987, *51*, 673
51. H.A. Scheraga, 'Protein Structure', Academic Press, New York, 1961
52. S.E. Harding, PhD Thesis, University of Leicester, 1980
53. S.E. Harding and A.J. Rowe, *Int. J. Biol. Macromol.*, 1982, *4*, 357
54. M. Rholam and P. Nicolas, *Biochemistry*, 1981, *20*, 5837
55. S.E. Harding, *Biochem. Soc. Trans.*, 1986, *14*, 857 and 1359

3
Aggregation and Elastic Moduli of Weakly Interacting Macromolecules

By D. M. Heyes and J. R. Melrose

DEPARTMENT OF CHEMISTRY, ROYAL HOLLOWAY AND BEDFORD NEW COLLEGE, UNIVERSITY OF LONDON, EGHAM, SURREY TW20 0EX, U.K.

1 INTRODUCTION

The aggregation of colloidal macromolecules has an important effect on the physical properties of suspensions. The elastic compressional and shear moduli, for example, are sensitive to the extent and nature of association of the macromolecules.[1-4] The modulus of the solvent is a constant in this context, and it can therefore be decoupled from the component due to the macromolecules. This constant is zero for the shear modulus on the time scales of the structural evolution of the macromolecules, $\tau_r \sim \sigma^3 \eta_s / k_B T$, where σ is the hard- core diameter of the macroparticle, η_s is the viscosity of the solvent, k_B is the Boltzmann constant, and T is the temperature.[5] Here we describe the association of particles in terms of physically meaningful clusters and discuss how aggregation of interacting particles can be described in terms of *percolation* critical exponents, and extend this formalism to the treatment of time- dependent elastic moduli.[6-9] More strongly interacting particles may not be described by percolation, rather by cluster- cluster or gelation models.[6]

2 THE TECHNIQUE

We use the computer simulation technique of Molecular Dynamics, MD, to generate the positions of interacting model macromolecules as they evolve in time.[10] (Although, as we are in thermodynamic equilibrium Monte Carlo or Brownian Dynamics could have been used.) We use the LJ potential, $\phi(r) = 4\epsilon((\sigma/r)^{12} - (\sigma/r)^6)$, to represent the interactions between our model molecules.[10] The MD technique moves all the particles simultaneously by small distances ($\ll \sigma$) corresponding to time intervals in a numerical solution of Newton's equations of motion. The velocities are thermal and the force on each particle arises from the net contribution from all surrounding molecules. The molecules are contained within periodically repeated cells each containing N particles. These, *periodic boundary conditions* are necessary because the computer can follow the trajectories of only a limited number of interacting particles. By periodically repeating particles in space we eliminate surfaces also. The MD simulations were performed on cubic unit cells of volume V containing $N = 108, 256$ and 864 particles, to assess the system- size dependence of the results. We use LJ reduced units throughout, *i.e.*, $k_B T/\epsilon \to T$, and number density, $p = N\sigma^3/V$. Unless stated distance is in LJ σ.

3 CLUSTER STATISTICS

We will describe a method for characterising the association of interacting particles by partitioning them into clusters. For this purpose a set of particles is considered to be part of the same cluster if each member is separated from at least one of the others by a distance $\leq \sigma_s$, which is arbitrary but is usually $\sim \sigma$, the core diameter of the particle. A percolating cluster, PC, is a cluster having infinite extent. Within the framework of the periodically repeating cells of Molecular Dynamics, a sufficient criterion for percolation is for a particle and its image to belong to the same cluster. The existence of a percolating cluster should have a pronounced effect on physical properties. It confers rigidity on the aggregate and it therefore appears at a density for which structural characterisation is important. The nature of the short range interactions is unimportant because the phenomenon is dominated by large scale or long range fluctuations.

For an infinite number of molecules there is a well- defined density, p_c, at which there is a finite probability of finding a percolating cluster.[6-8] Below this p no percolating cluster is found. For the small N systems considered here this transition is not sharp but continuous. As p increases the probability of discovering a

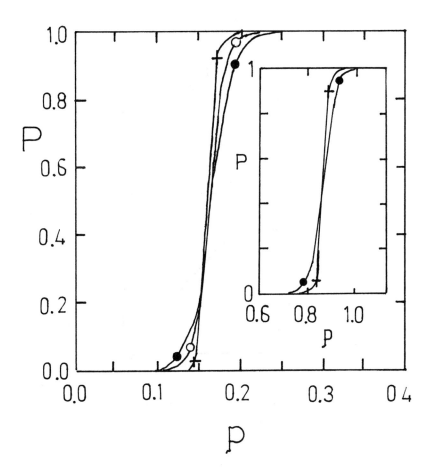

Figure 1

The percolation fraction, P, against density, p for $T = 1.4562$, $\sigma_s = 1.4688$. key: $N = 108, \bullet$, $N = 256, \circ$, and $N = 864, +$. The insert is for $\sigma_s = 1.02816$.

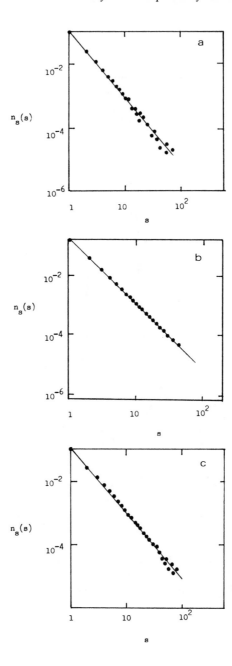

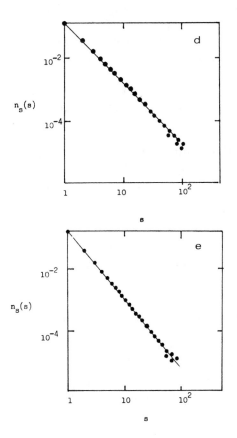

Figure 2

The cluster number distribution for non- percolating clusters, $n_s(s)$ for the LJ state points: (a) $p = 0.16433, (p = p_c), T = 1.4562, \sigma_S = 1.4688$ and $N = 864$. $\tau = 2.0 \pm 0.1$. (b) $p = 1.2564, (p = p_c), T = 6.0, \sigma_S = 0.92226$ and $N = 864$. $\tau = 2.2 \pm 0.1$. (c) $p = 0.2585, (p = p_c), T = 6.0, \sigma_S = 1.31751$ and $N = 864$. $\tau = 2.2 \pm 0.1$. (d) $p = 0.2415, (p \ll p_c), T = 6.0, P = 0.072, \sigma_S = 1.31751$ and $N = 864$. $\tau = 1.9 \pm 0.1$. (e) $p = 0.86821, (p = p_c), T = 1.4562, \sigma_S = 1.02816$ and $N = 864$. $\tau = 2.2 \pm 0.1$.

percolating cluster at a time- step increases smoothly from 0 to 1. To define a p_c for a particular N we make use of the density at which the probability of discovering a percolating cluster in a time step - the percolation fraction, P - equals 0.5.[11] This is because the density at which $P = 0.5$ shows the least system size dependence, as shown in Figure 1. The $P(p)$ curves become sharper with increasing N with $p_c(N)$ moving to lower p as $\sim N^{1/3\nu}$, consistent with a recent value for $\nu = 0.88 \pm 0.06$ for random percolation .[8,11] The value of p_c for the LJ fluid is determined principally by the volume fraction of the outer search spheres, $\phi_s = (\pi/6)p_s$, where $p_s = N\sigma_s^3/V$, modified by the presence of an effective hard- core of diameter, σ_{HS}, where $\sigma_{HS} \leq \sigma_s$.[12] For macromolecules with $\sigma_s/\sigma_{HS} \geq 1.2$, the percolation threshold occurs at $\phi_s = 0.35 \pm 0.02$. As this ratio, $\sigma_s/\sigma_{HS} \to 1$, ϕ_s increases to 0.64, because the hard- cores start to hinder the overlap of the soft- cores. An effective hard- core can be attached to soft- particles, such as the LJ molecule to demonstrate this.[12,13] We refer to $\sigma_s/\sigma_{HS} \to 1$ as the hard- core limit and $\sigma_s/\sigma_{HS} \to \infty$ as the soft- core limit.

The distribution of different sized clusters is characterised by the cluster number distribution function, n_s, which is the time-average number of clusters containing s particles, N_s divided by N, i.e., $n_s = N_s/N$.[14,15] For finite periodic systems there is an upper bound on s, i.e., $1 \leq s \leq N$, which will distort the behaviour of n_s for large s. This becomes less important in practice in the $N \to \infty$ limit.

For non- interacting particles thrown randomly on a 3-dimensional lattice of arbitrary symmetry ("random lattice percolation"), then it has been discovered that at the percolation threshold,[8,14]

$$n_s(p_c) \propto s^{-\tau}. \tag{1}$$

where $\tau = 2.2$ is followed.[8,14] Figure 2 gives some typical $n_s(p_c)$ and $n_s(p)$, demonstrating that this relationship is obeyed well, with no statistically significant temperature dependence of $\tau = 2.1 \pm 0.1$ at $T = 1.4562$ and $T = 6.0$. Random lattice studies suggest that close to p_c,

$$n_s(p) = n_s(p_c)f(z), z \equiv (p - p_c)s^\sigma; s \to \infty, p \to p_c. \tag{2}$$

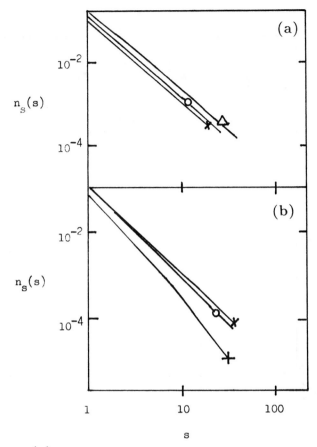

Figure 3 (a)

The non- percolating n_s for three $p \leq p_c$ states. $T = 6.0, \sigma_s = 1.31751, N = 256$. key: $p = 0.26093, \times$, (percolation threshold for this N), $p = 0.24787, \circ, p = 0.212517, \triangle$

(b)

As for figure 3(a) except that three $p \geq p_c$ states are considered. key: $p = 0.26093, \times$, (percolation threshold for this N), $p = 0.27398, \circ$ and $p = 0.30206, +$.

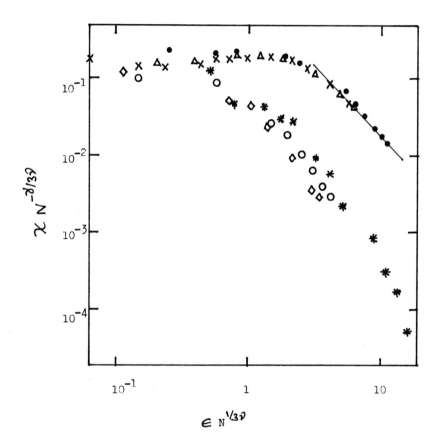

Figure 4

The susceptibility, χ for $T = 1.4562, \sigma_s = 1.4688; \gamma = 1.8\pm$ 0.1 from the upper curve which is above p_c. key: $p \leq p_c$, upper curve:$N = 108, \times, N = 256, \triangle, N = 864, \bullet.$ $p \geq p_c$, lower data points: $N = 108, \diamond, N = 256, \circ, N = 864, \star.$

where $f(z)$ is a universal function and $\sigma = 0.45$.[8] Figure 3 gives typical examples. As is revealed there, the $n_s(p)$ in the vicinity of p_c are very similar in form to $n_s(p_c)$, manifesting a change in the prefactor A only. The form of $f(z)$ is such that descending just below p_c, $n_s(s)$ **increases** for small s but decreases for large s. As $(p - p_c)$ becomes more negative then this change in behaviour moves progressively to smaller s.

Above p_c in $n_s(p)$ there are always fewer clusters of all sizes s than at p_c, a trend which becomes more pronounced as $(p - p_c)$ becomes more positive, *i.e.*, $n_s(p) \leq n_s(p_c)$.

The susceptibility , χ, is,

$$\chi = \sum_s{}' s^2 n_s(p). \tag{3}$$

The "*'*" denotes the omission of the largest cluster at each sample configuration. The largest cluster discovered each time step is either a percolating cluster, should (at least) one exist, or the largest non- percolating cluster. For non- interacting particles thrown randomly on a 3- dimensional lattice of arbitrary symmetry, then it has been discovered that about p_c,

$$\chi = c_- \mid p - p_c \mid^{-\gamma} . \tag{4}$$

for $p \leq p_c$ and,

$$\chi = c_+ \mid p - p_c \mid^{-\gamma} . \tag{5}$$

for $p \geq p_c$. where $\gamma = 1.78 \pm 0.06$.[8,11] The amplitude ratio, c_-/c_+ is ~ 10, [15] indicating that the 'mean' size of the the clusters is somewhat larger below p_c than above p_c for the same $\mid p - p_c \mid$. An example is given in Figure 4.

We have evaluated χ at a range of p about p_c. We find for example, in figure 4, for a $T = 1.4562$ soft- core state that, $\chi \propto \mid p - p_c \mid^{-\gamma}$ for $p \leq p_c$ with $\gamma = 1.8 \pm 0.1$ in agreement with the value for 3D lattices, *i.e.*, $\gamma = 1.8$.[8] The different N values collapse onto the same curve when finite- scaling corrections are applied.[7,8] We find, however, that in the hard- core limit, scaling

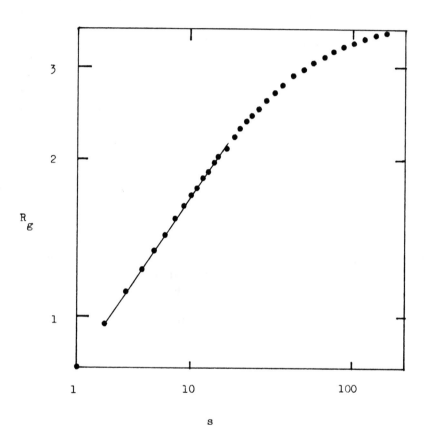

Figure 5

The s- dependence of the radius of gyration, R_g, for the LJ state point , $p = 0.26093$, •, $\sigma_s = 1.31751$, $N = 256$, $T = 6.0$. For $R_g \propto s^{1/D_f}$, then the line in the figure indicates $D_f = 2.3 \pm 0.1$

is not obeyed. Therefore, for thermally equilibrated systems there is a breakdown of universality at high density when $\sigma_s/\sigma_{HS} \to 1$.

The amplitude ratios indicate that clusters are smaller (*i.e.*, contain a smaller number of particles) above p_c than below it for the same $| p - p_c |$. This effect is more pronounced in lattices than for continuum systems. However, we observe another effect that reduces the size of the non- percolating clusters above p_c, which is a finite N effect. Above p_c there is a departure from universality. The different N curves do not superimpose, and an effective $\gamma \approx 3.3$ can be defined. The non- percolating clusters decrease more dramatically in size above p_c with increasing $| p - p_c |$ because the percolating cluster has a greater chance of incorporating any non- percolating clusters, as $1 \leq s \leq N$.

We now investigate the extent of openness of the non- percolating clusters at the percolation threshold by considering their radius of gyration, R_g, [6],

$$R_g = \frac{1}{2} < \sum_i^{s-1} \sum_{j \neq i}^{s} \underline{R}_{ij}^2 / s(s-1) >^{1/2} . \tag{6}$$

where $\underline{R}_{ij}$ is the vector separation between particles i and j. The scaling relationship here is $R_g \propto s^{1/D_f}$ as $s \to \infty$, and where D_f is the fractal or Hausdorff dimension.[6] Figure 5 gives a typical example for supercritical LJ states at the percolation threshold. There is an intermediate s regime where this scaling relationship is obeyed, with $D_f = 2.3 \pm 0.10$, slightly lower than the accepted value for 3D static random lattice percolation, $D_f = 2.5$.[7] In practice, this scaling relationship applies well for $0.2 \leq P \leq 0.8$. Therefore the fractal dimension of the percolating clusters is insensitive to p about p_c. As exhibited by n_s, the finite size of the MD cell leads to deviations from scaling laws in the $s \to N$ limit.

The pair radial distribution function, $g(r)$ and pair connectedness function, $p(r)$, for pair separations, r, are probes of the local structure in the whole fluid and in the percolating clusters, respectively.[11,16] They are formally very similar.

$$g(r) = n(r)/(4\pi r^2 \rho \delta r). \tag{7}$$

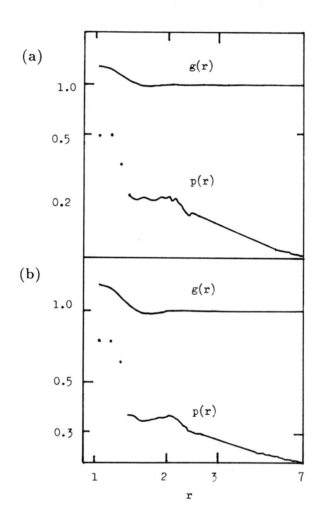

Figure 6

Log-log plot of $g(r)$ and $p(r)$ for: (a) $p = 0.2415, (p \ll p_c), T = 6.0, N = 864, P = 0.07, \sigma_s = 1.31751, D_f = 2.4 \pm 0.1.$ (b)$p = 0.2585(p_c), T = 6.0, N = 864, P = 0.41, \sigma_s = 1.31751, D_f = 2.7 \pm 0.1.$

where δr is the radial increment for $n(r)$; $n(r)$ is the number of particles found on average within $r - \delta r/2 \leq r \leq r + \delta r/2$.

$$p(r) = n(r)P_\infty/(4\pi r^2 \rho \delta r). \tag{8}$$

The search for pairs in $p(r)$ is restricted to those particles within the same PC. As $r \to \infty$ then $p(r) \to P_\infty^2$. For finite r there is a regime in which $p(r) \sim r^{D_f - 3}$. For the small N considered here it is not possible to go out far enough in r to determine D_f in the hard- core limit. The $p(r)$ look similar to the $g(r)$ but attain a lower limiting value. When the pair separation becomes comparable to L, the sidelength of the simulation cell, the dimension of PC must approach the dimension of the space, d (=3 here). This is a finite- size artefact. In the hard- core limit the $p(r)$ does not go much beyond the first two peaks in $g(r)$ before $r \approx L/2$. Therefore the finite size of the cell precludes us from observing long- range correlations of the particles. There is not sufficient r range to extract a D_f. We therefore confine our attention to the soft- core limit, where percolation takes place at lower density and therefore the radius range is greater ($\leq L/2$). The $g(r)$ and $p(r)$ for two $T = 1.31751$ soft- core states are given in Figure 6. The low density state in Figure 6(a), $p = 0.2415$ coincides with $P = 0.07$ and is therefore below p_c. The D_f is 2.4 ± 0.1. The Figure 6(b) examines the PC at p_c. There is an intermediate r range with $D_f = 2.7 \pm 0.1$ before the PC becomes Euclidean as the distance scale of the cluster becomes comparable to the size of the MD cell. Therefore the dimension of the PC approaches the Euclidean dimension (= 3), here with increasing p.

4 ELASTIC MODULI

We now turn our attention to the physical behaviour of the clusters close to p_c as probed by the elastic moduli. In the hard- core (*i.e.*, dense suspension limit) the Zwanzig-Mountain expressions for the infinite- frequency moduli can be used, because bond lengthening interactions dominate during strain.[17−19] For the infinite- frequency shear modulus, G_∞,

$$G_\infty = pk_BT + \frac{2\pi p^2}{15} \int_0^\infty dr g(r) \frac{d}{dr}(r^4 \frac{d\phi}{dr}). \tag{9}$$

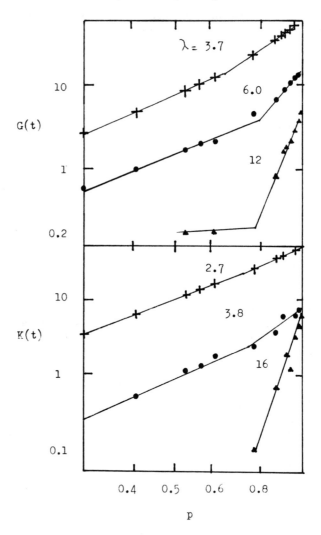

Figure 7(a)

log-log plot of $G(t)$ <u>versus</u> p for $T = 1.4562$ Lennard-Jones states $N = 256$. key: $t = 0.0, +, t = 0.1, \bullet, t = 0.3, \blacktriangle$.

(b)

as for Figure 7(a), except $K(t)$ is plotted.

where $g(r)$ is the radial distribution function. For the infinite-frequency compressional modulus, K_∞,

$$K_\infty = \frac{2}{3}pk_BT + P_{osm} + \frac{2\pi p^2}{9} \int\limits_0^\infty dr\, g(r)r^3 \frac{d}{dr}\left(r\frac{d\phi}{dr}\right). \qquad (10)$$

where P_{osm} is the osmotic pressure, given by,

$$P_{osm} = pk_BT - \frac{2\pi p^2}{3} \int\limits_0^\infty dr\, g(r)r^3 \frac{d\phi}{dr}. \qquad (11)$$

where $g(r)$ is the radial distribution function. This central-force approach is inadequate in the soft- core regime ($i.e.$, $\sigma_s/\sigma_{HS} \geq$ 1.2) appropriate to weakly aggregating dispersions. The percolation transition takes place at small number densities causing bond- bending to be more important than bond- contraction or extension. Equations (9)-(11) are clearly inadequate as they predict a finite G_∞ and K_∞ in the limit $p \to 0$, whereas the modulus of a floc vanishes below a critical p, p_m (which in the case of strong ('stiff- spring') interactions equals p_c but for weak-spring interactions equals dp_c, where d is the dimension of the space.[3]) It is not easy to relate the Lennard-Jones p_m to p_c because there is no single distance scale ($i.e.$, σ_s) that characterises G_∞ or K_∞

The exponents are for $G_\infty, K_\infty \propto (p - p_m)^\lambda$, where $\lambda =$ 3.75 $\pm$ 0.5.[1,20] The exponent near p_m depends on the physical property. For example, the exponent for conductivity is 1.9 $\pm$ 0.1.[21] A problem with lattice-based studies of these moduli is that they do not recognise that real aggregates are flowing objects and are only 'solid-like' on short time-scales. In any measurement of the moduli it is important to allow for structural relaxation after the application of the strain. The longer the delay, the higher p_m is. Indeed the relaxation times diverge at p_c. This is is illustrated in Figure 7 which shows the p- dependence of $G(t)$ (Figure 7(a)) and $K(t)$, (Figure 7(b)) obtained from Green-Kubo time correlation functions.[23] Note that $G(0) = G_\infty$ and $K(0) = K_\infty - K_0$, where K_0 is the isothermal thermodynamic compressional modulus. The shear modulus measured instantaneously after the strain is applied ($i.e.$, $G(0)$) and after a period of relaxation show a shift of the p_m to higher p and greater λ as t increases. The comparable curves for the bulk moduli behave similarly. To be sure of extracting G_∞ for a dispersion the experimental timescale, $t \ll \tau_r$ so that little movement within the first coordination shell is allowed to take place. For $1\mu m$ particles in water $\tau_r \sim 1$ sec.[5]

Acknowledgements

D.M.H. gratefully thanks *The Royal Society* for the award of a *Royal Society 1983 University Research Fellowship.* J.R.M. thanks the S.E.R.C. for the award of a post-doctoral research fellowship. We thank the S.E.R.C. for the award of computer time at the University of London Computer Centre, and the RHBNC Computer Centre for use of their computing facilities.

References

1. I. Webman, in 'Fractals in Physics', eds. L. Pietronero and E. Tosatti, Elsevier, 1986, page 343.

2. S. Feng, M.F. Thorpe and E. Garboczi, *Phys. Rev. B*, 1985, **31**, 276.

3. W. Tang and M.F. Thorpe, *Phys. Rev. B*, 1988, **37**, 5539.

4. L.C. Allen, B. Golding and W.H. Haemmerle, *Phys. Rev. B*, 1988, **37**, 3710.

5. D.M. Heyes, *J. Non-Newt. Fluid Mech.*, 1988, **27**, 47.

6. H.J. Herrmann, *Phys. Rep.*, 1986, **136**, 153.

7. D. Stauffer, 1986 *On Growth and Form* ed H.E. Stanley and N. Ostwesky (Martinus Nijhoff) p 79.

8. D. Stauffer, *Phys. Rep.*, 1979, **54**, 1.

9. I. Balberg and N. Binenbaum, 1987, *Phys. Rev. A*, **35**, 5174.

10. D.M. Heyes, 1987 *J. Chem. Soc., Faraday Trans. II*, 1987, **83**, 1985.

11. N.A. Seaton, and E.D. Glandt, *J. Chem. Phys.*, 1987, **86**, 4668.

12. A.L.R. Bug, S.A. Safran, G.S. Grest, and I. Webman, *Phys. Rev. Lett.*, 1985, **55**, 1896.

13. D.M. Heyes, and J.R. Melrose, *Mol. Sim.*, 1988, in press.

14. J.P. Lu, and J.L. Birman, *J. Stat. Phys.*, **46**, 1987, 1057.

15. I. Balberg, *Phys. Rev. B*, 1988, **37**, 2391.

16. E.M. Sevick, P.A. Monson, and J.M. Ottino, *J. Chem. Phys.*, 1988, **88**, 1198.

17. R. Zwanzig and R.D. Mountain, *J. Chem. Phys.*, **43**, 1965, 4464.

18. M.J. Grimson, *Mol. Phys.*, 1986, **59**, 737.

19. R. Kesavamoorthy and A. K. Arora, *J. Phys. C: Solid State Phys.*, 1986, **19**, 2833.

20. R. Buscall, P.D.A. Mills, J.W. Goodwin and D.W. Lawson, *J. Chem. Soc., Faraday Trans. I*, 1988, 84, in press.

21. N.A. Seaton and E.D. Glandt, *J. Phys. A: Math. Gen.*, 1988, **20**, 3029.

22. B.K. Chakrabarti, D. Chowdhury, and D. Stauffer, *Z. Phys. B*, 1986, **62**, 343.

23. K.D. Hammonds and D.M. Heyes, . *J. Chem. Soc., Faraday Trans. II*, 1988, in press.

4

Biological Applications of Solid State NMR

By M. A. Hemminga and T. W. Poile

DEPARTMENT OF MOLECULAR PHYSICS, AGRICULTURAL UNIVERSITY, DREIJENLAAN 3, 6703 HA WAGENINGEN, THE NETHERLANDS.

1 INTRODUCTION

During the past decade, solid state NMR has become a powerful tool for studying many biological systems. This is not only because commercial NMR spectrometers have become available, but also because remarkable developments in the theory of high-power NMR as applied to solid systems have been made. A large number of papers about biological applications of solid state NMR has appeared in the literature covering systems such as crystals and powders of biological macromolecules, solutions of supra-molecular systems and anisotropic systems such as biological and model membranes.[1-4]

It will not be the purpose of this paper to review all the biological applications of solid state NMR. An overview will be given of the basic interactions that give rise to NMR spectra and the instrumental features of high-power solid state NMR. This will be illustrated with examples from studies of phospholipid membranes and viral coat proteins.

2 BASIC PRINCIPLES

Chemical Shift Anisotropy

A nucleus in a molecule in the presence of an external magnetic field senses a magnetic shielding effect due to the surrounding electrons. This reduction of the externally applied magnetic field leads to a characteristic resonance

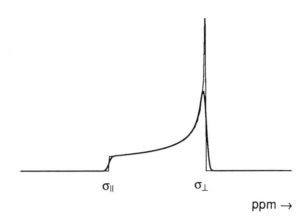

$\sigma_{\parallel}$ $\sigma_{\perp}$

ppm →

<u>Figure 1</u> Schematic powder line shapes of a nucleus with an axially symmetric chemical shift tensor σ with components $\sigma_{\parallel}$ and $\sigma_{\perp}$. The unbroadened line shape is given by Eqn. 12, the broadened line shape by Eqn. 9.

position for the nucleus in an NMR spectrum. The change in resonance position is known as the chemical shift and is proportional to the extent of shielding, σ. Chemical shift is measured as a proportion of the applied field in units of ppm. As each nucleus in a molecule is shielded to differing extents this effect is used for the finger printing of organic and biological molecules in solution. For solids, however, the magnitude of the chemical shift can depend on the orientation of the molecule with respect to the magnetic field, giving chemical shift anisotropy effects. Chemical shift anisotropy is especially large for nuclei in molecular groups with an anisotropic electron distribution, i.e. [13]C nuclei in C=O and C=C bonds, and ^{31}P nuclei in phosphate groups.

Mathematically such an anisotropic interaction is best described by a tensor notation. This will be illustrated for a ^{13}C nucleus in a C=C chemical group. Because of the axial symmetry of this chemical bond, the chemical shift tensor σ contains two components, $\sigma_{\parallel}$ and $\sigma_{\perp}$. This means that when the external magnetic field B is oriented along the ‖ axis, the NMR resonance of the ^{13}C nucleus appears at a magnetic field value $B_{\parallel} = (1 - \sigma_{\parallel})B$, whereas

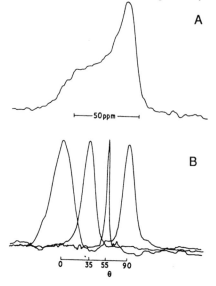

Figure 2 129-MHz ^{31}P NMR spectra of non-oriented (A) and oriented (B) dimyristoyl phosphatidylcholine bilayers in the liquid crystalline phase (30 °C). The angle θ is between the normal to the bilayers and the magnetic field.[6]

when the magnetic field B is oriented along the ⊥ axis, the NMR resonance of the ^{13}C nucleus appears at a magnetic field value $B_\perp = (1 - \sigma_\perp)B$. This situation applies for a single crystal system, where the magnetic field can be oriented in various ways with respect to the molecule. For molecules in a powder, all possible orientations with respect to the magnetic field are found and a so-called powder line shape is produced (see Figure 1). Because of line-broadening processes a broadened line shape is obtained experimentally.[5]

As an example, such a line shape is shown in Figure 2, which is the ^{31}P NMR spectrum obtained from randomly oriented dimyristoyl phosphatidyl-choline bilayers in the liquid crystalline phase. For comparison the oriented spectra as a function of the angle θ between the normal to the bilayers, which is identical with the symmetry axis of the tensor, and the magnetic field are also shown.[6] This angular dependence of the chemical shift value σ(θ) is given by

$$\sigma(\theta) = \frac{1}{3}(\sigma_\parallel - \sigma_\perp)\,(3\cos^2\theta - 1). \qquad\qquad [1]$$

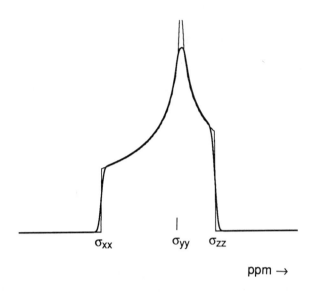

σ_{xx} σ_{yy} σ_{zz}

ppm →

Figure 3 Schematic powder line shapes of a nucleus with a non-axially symmetric chemical shift tensor σ with components σ_{xx}, σ_{yy}, and σ_{zz}.

For a non-axially symmetric chemical bond, the chemical shift tensor σ consists of three components σ_{xx}, σ_{yy}, and σ_{zz}, where x, y, and z are the principal axes of the tensor. The corresponding unbroadened and broadened powder NMR line shapes are illustrated in Figure 3. The width of the NMR spectra arising from anisotropic chemical shift effects depend on the compound. These are about 200 ppm for both ^{13}C nuclei[7] and ^{31}P nuclei in phospholipid molecules.[8] The corresponding frequency range is 15 kHz and 25 kHz, respectively, on a 300 MHz NMR spectrometer. For phospholipids the chemical shift anisotropy has been used to provide information on the phase adopted by the molecules (lipid polymorphism).[9]

Order parameters

In a fluid, in which the molecules are undergoing rapid isotropic motion, the anisotropic interactions completely average out. In the case of chemical shift anisotropy the NMR resonance appears then at the isotropic chemical shift position given by $\sigma_{iso} = \frac{1}{3}(\sigma_{xx} + \sigma_{yy} + \sigma_{zz})$.

Incomplete molecular averaging is found in systems that are partially oriented (i.e. have crystal and fluid-like properties) or in systems where anisotropic rotations take place. For example, in biological membranes the lipid molecules rotate preferentially about their long molecular axes. The rigid lattice chemical shift tensor σ is then reduced to a time-averaged tensor σ_{av}, which is axially symmetric. If an x, y, z, coordinate system is attached to the molecule, the average fluctuations of these axes around a preferential director are described by a set of order parameters, one for each axis:

$$S_{ii} = \tfrac{1}{2}<3\cos^2\theta_i - 1> \quad (i=1,2,3=x,y,z), \qquad [2]$$

where the brackets represent a time average of the angular fluctuations. Since

$$\sum_{i=1}^{3} S_{ii} = 0, \qquad [3]$$

only two order parameters are independent. If the effective molecular fixed tensor is axially symmetric about the z-axis then only S_{33} needs be specified. This order parameter is given by

$$S_{33} = \tfrac{1}{2}<3\cos^2\theta - 1>, \qquad [4]$$

where θ is the angle between the z-axis and the director.

As each order parameter is a measure of the time averaged fluctuations of the nucleus being considered, it will be affected by both local conformational changes in the molecule and the motional properties of the molecule itself. Thus for a lipid molecule in a bilayer such motional effects can be intramolecular but may also be related to the vesicle tumbling rate[10] or the lateral diffusion rate of the lipid molecule in the bilayer.[11]

Electric Quadrupole Interactions

Nuclei, which have a nuclear spin quantum number $\tfrac{1}{2}$, can be considered as having a spherical shape. This is the case for nuclei such as ^1H, ^{13}C, ^{15}N, and ^{31}P. However, nuclei with nuclear spin quantum numbers ≥ 1 are non-spherical and possess a quadrupole moment. The interaction of the nuclear quadrupole moment with the surrounding electric field gradient of

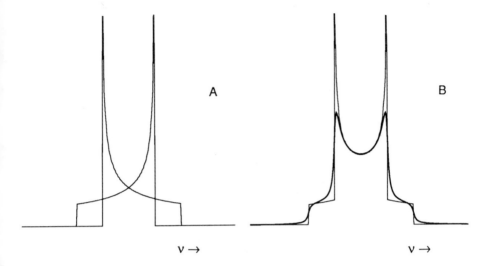

Figure 4 Schematic line shapes of a spin 1 nucleus with an axially symmetric quadrupolar interaction. (A) Unbroadened powder line shapes for each of the transitions. (B) Complete powder lineshape with and without linebroadening.

the bonding electrons is known as quadrupolar coupling, which can give rise to very large splittings in the NMR spectra. A well-studied nucleus is the ^{2}H nucleus, a spin 1 nucleus, which is often substituted for protons and used as a non-perturbing probe in the NMR spectroscopy of biological systems such as phospholipids and proteins.

For a single crystal, two resonance lines are observed of spin 1 nuclei with a frequency spacing known as the quadrupole splitting Δv_Q. For an axially symmetric quadrupolar tensor this splitting is given by

$$\Delta v_Q = \frac{3}{4} \frac{e^2 q Q}{h} S_{33} (3 \cos^2\theta - 1), \qquad [5]$$

where e^2qQ/h is the quadrupolar coupling constant, S_{33} is an order parameter (Eqn. 4) and θ is the angle between the symmetry axis and the applied magnetic field. The quadrupolar coupling constant e^2qQ/h has a value of ≈ 170 kHz for a C-^{2}H bond.[12] From Eqn. 5 it can be seen that the angular dependence of the resonance position depends on $3 \cos^2\theta - 1$ as does the chemical shift interaction (Eqn. 1). Since the quadrupole interaction produces two resonance lines rather than one, the powder line shape that is obtained consists of two overlapping powder line shapes from Figure 1. This is illustrated in Figure 4.

Dipolar Interactions

In a molecule generally a large number of nuclei with magnetic properties are present. Between these nuclei interactions arise, with each nucleus sensing the magnetic moments of the surrounding nuclei. Such interactions are called dipolar interactions, and can be described also in terms of a tensor notation. Dipolar couplings are very strong and splittings can be in the order of 100 kHz. For an axially symmetric dipolar coupling tensor, a splitting is obtained that has an angular dependence similar to that obtained for quadrupolar interactions. Therefore the line shapes for dipolar coupled nuclei are similar to those shown in Figure 4. In practice, the line shapes of solids are much more complex than this, because one nucleus has interactions with many other nuclei. However, an interesting property of this type of inter-nuclear interaction is that it is determined by the geometric properties of the molecule. The strength of the dipolar interaction between two nuclei is dependent on the angle between the magnetic field and the axis linking the two nuclei and is also proportional to $1/r^3$ where r is the distance between the two nuclei. These properties may therefore be deduced from an analysis of these dipolar interactions, for example if suitable isotope substitutions have been carried out, so that specific information can be obtained.

The effect of ^1H dipolar couplings can be well observed in the angular dependent line widths in the ^{31}P NMR spectra of oriented lipids in Figure 2. The line widths depend on θ through a $3 \cos^2\theta -1$ dependence, as may be expected for an axially symmetric system.

Relaxation Effects

Motional effects in spin systems lead to modulations of the NMR interactions. Depending on the time scale of the motions different effects on the solid state NMR spectra can be observed. If the motion is slow (i.e. the correlation time τ_c of the motion is long), an additional broadening of the spectral lines is obtained, given by $\Delta\omega \approx \tau_c^{-1}$. If the motion is very fast relative to the splittings in the NMR spectra, the interaction tensors can be averaged first to give the line positions. The modulations by the remaining anisotropies of the interactions determine the relaxation of the nuclear spins. Since motional averaging results in Lorentzian shapes, the line width at half height

$\Delta v_{1/2}$ is given by $\Delta v_{1/2} = 1/\pi T_2$, where T_2 is the spin-spin relaxation time. The relaxation time T_2 is given by the approximation

$$1/T_2 \approx \Delta^2 \tau_c, \qquad [6]$$

where Δ is the reduction of the splitting in the non-averaged NMR spectrum as compared to the averaged spectrum.

Another source of dynamic information from an NMR experiment is the spin-lattice relaxation time T_1. This relaxation time is related to the recovery of magnetization of the nuclear spins in the external magnetic field B_0. In isotropic fluids the expression of T_1 in terms of the motional correlation times is straightforward,[13] but for solids and anisotropic systems the equations become very complex. As a rule of thumb, it can be assumed that T_1 is determined by the spectral density at frequency $\omega_0 = \gamma B_0$, given by

$$1/T_1 \approx \Delta^2 \frac{\tau_c}{1+\omega_0^2\tau_c^2}. \qquad [7]$$

This shows that motions with frequencies close to ω_0 contribute to T_1.

In solid state NMR it is also possible to lock the magnetization of the nuclei along the magnetic field component B_1 of the rf field in the rotating frame. The relaxation time $(T_{1\rho})$ in this frame of reference governed by B_1 can then be studied. $T_{1\rho}$ is given by an equation similar to Eqn. 7, but with ω_0 replaced by $\omega_1 = \gamma B_1$. This opens the possibility of studying motions in a much lower frequency range, since $B_1 \approx B_0/1000$.

3 SPECTROSCOPIC TECHNIQUES

Pulse Width and Power

One of the characteristics of solid state NMR spectra is that the spectral width is very large. Exciting such broad line shapes is possible only if the width of the rf pulses is very short. This may be seen from a Fourier analysis of a pulse of rf radiation with duration τ_p and angular frequency ω_0, which gives an excitation amplitude distribution given by $\sin\{\frac{1}{2}(\omega-\omega_0)\tau_p\}/\frac{1}{2}(\omega-\omega_0)$. If it is assumed that a range of 1 MHz has to be excited with a maximum loss of 10%

in excitation amplitude, the value of τ_p is ≈ 3 µs. The rotation θ of the magnetization during the pulse is $\theta = \gamma B_1 \tau_p$. To give a 90° pulse, B_1 has to be ≈ 1 mT for ^{1}H nuclei at 300 MHz. This is a very high value that only can be achieved using high power (i.e. kW) transmitters. High power is not only necessary for exciting nuclei, but is also required for decoupling nuclei.

High-Power Decoupling

In some NMR experiments one is not interested in the dipolar couplings. This is the case if the dipolar interactions are large and contribute in a substantial way to the line shape so that it is very difficult to analyse the spectra. It is then possible to decouple the dipolar interactions by broad-band high-power irradiation on the non-observable nuclei, thus leading to a simplification of the NMR spectra. This is particular useful for recording ^{13}C and ^{31}P NMR spectra free of dipolar interactions with ^{1}H nuclei by ^{1}H-decoupling.

Cross-Polarization

A method to increase the sensitivity of an NMR experiment is cross-polarization. In this case the magnetization of a highly abundant nucleus I is transferred to a less abundant nucleus S. The increase of magnetization is theoretically given by γ_I / γ_S. This principle is employed for recording ^{13}C or ^{31}P NMR spectra where the magnetization of the ^{1}H nuclei can be used to increase the sensitivity. This is called proton-enhancement.

Magic-Angle-Spinning NMR

Another method to simplify NMR spectra of solids is to use the technique of magic-angle-spinning (MAS). The material is enclosed in an air turbine (spinner), which rapidly rotates about an axis that makes an angle ε with respect to the external magnetic field. This method relies on the fact that anisotropic parts of the chemical shift interactions are completely removed if the angle ε is the magic angle (54° 44'), so that only the isotropic chemical shift positions are obtained. This method has been shown to be very useful for ^{13}C and ^{31}P nuclei. Also the dipolar couplings are suppressed by MAS, but to fully reduce the dipolar effects, ^{1}H-decoupling has to be used as well. The MAS NMR spectra, however, still contain side-bands arising from a modulation of the anisotropic chemical shift by the spinning process. If these

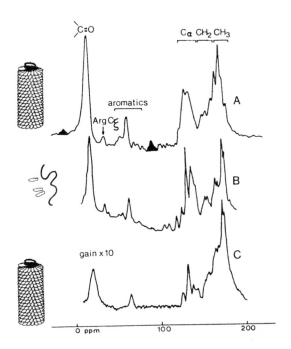

Figure 5 Comparison of conventional and MAS ^{13}C NMR spectra of tobacco mosaic virus (TMV). (A) 45-MHz proton-enhanced and proton decoupled MAS ^{13}C NMR spectrum of intact TMV. The black peaks indicate spinning side bands of the carbonyl resonance and the main aromatic resonance. (B) 90-MHz broad-band proton-decoupled ^{13}C NMR spectrum of dissociated TMV at pH 11. (C) 90-MHz broad-band proton-decoupled ^{13}C NMR spectrum of intact TMV at pH 7.2. In (B) and (C) TMV is 12% ^{13}C-enriched.[16]

side-bands overlap with interesting resonances, a different spinning speed can be chosen, or side band suppression can be applied.[14,15]

It would be expected that MAS NMR should give fluid-like spectra from solids, but this is not generally true. In solids there is always a distribution of isotropic chemical shifts which leads to a broadening of the NMR lines. In fluids these effects are not visible, since they are averaged out. As an example, Figure 5 shows ^{1}H-decoupled MAS ^{13}C NMR spectra of an intact tobacco mosaic virus particle, compared with the spectrum of the dissociated virus.[16] Line broadening effects are a drawback of MAS NMR, but by suitable isotope substitutions and the application of various spectroscopic techniques,[17,18] these problems may be overcome .

Spin Echo Techniques

A problem with solid state spectra is that the decay of the NMR signal can be very fast and so can be adversely affected by receiver recovery. This may be overcome by the use of an echo pulse sequence. For spin 1 nuclei such as ^{2}H nuclei the quadrupole echo sequence is used ($90°_x$-τ-$90°_y$-τ-).[19] The first pulse generates a normal free induction decay and the second pulse refocuses it with an echo occurring at time 2τ. If τ is chosen so that it is longer than the time taken in which it takes the receiver to recover from the first pulse then an undistorted spectrum can be obtained though with some loss in signal intensity. This loss in intensity is related to the spin-spin relaxation time T_{2e}. By using a series of echo delays and measuring the decay of the echo with time this relaxation time can be determined.

To use this technique properly the free induction decay must be adequately digitised so that the top of the echo can be accurately determined. Therefore a fast transient recorder is a necessity. This digitisation also enables the large spectral widths encountered in solid state NMR to be examined. Care must be taken to ensure that the experiment is on resonance and that the 90° pulse widths and phases are accurately set. To overcome pulse inaccuracies phase cycling can be introduced.[3]

For spin $\frac{1}{2}$ nuclei the Hahn echo sequence ($90°_x$-τ-$180°_y$-τ-) can be used to refocus the NMR signal overcoming the same instrumental problems as the quadrupole echo sequence.[20]

Two-Dimensional NMR

Two-dimensional techniques have already shown their great impact in resolving complex spectra in high-resolution NMR by spreading out the information along two frequency axes. Also in solid state NMR these techniques have shown to be very powerful in extracting useful information from NMR spectra (see Morden and Opella, this book). Two-dimensional ^{2}H NMR has been used to obtain information on slow molecular motions in solids.[21]

4 LINE SHAPE ANALYSIS

Computer Simulation

Analysis of the NMR line shapes of solids is in most cases directly possible, as is illustrated in Figures 1,3 and 4. However, in many cases such an analysis is not possible, so that more powerful methods must be employed. One such method is to simulate the spectrum by computer. By comparing such simulations with experimental data it is possible to obtain further information about the tensor interactions affecting each spectrum.

For all the interactions discussed if the tensor interaction is axially symmetric then the orientation dependence of the resonance position is dependent on the term $3 \cos^2 \theta -1$. If

$$\xi = \tfrac{1}{2}(3 \cos^2 \theta -1), \tag{8}$$

then a general equation for the simulation of an axially symmetric powder lineshape is given by

$$f(v) = \int_{-1/2}^{1} p(\xi)\, g(v,v(\xi))\, d\xi , \tag{9}$$

where $p(\xi)$ is the probability density for the lineshape associated with the frequency ξ and $g(v,v(\xi))$ is the function that describes that lineshape. Two types of lineshape function can be considered which are Lorentzian and Gaussian lineshapes. A resonance $g(v,v(\xi))$ with a Gaussian lineshape is described by

$$g(v,v(\xi)) = \frac{T^*_2}{\sqrt{2\pi}} \exp\{-2(\pi\, T^*_2\, (v -v(\xi)))^2\}, \tag{10}$$

while the Lorentzian lineshape is described by

$$g(v,v(\xi)) = \frac{1}{T_2^{*-2} + 4\pi^2(v -v(\xi))^2} . \tag{11}$$

Dynamic Properties of Biomolecular Assemblies

T^*_2 is the spin-spin relaxation time including the contribution from residual static interactions (mostly arising from dipolar effects) and magnetic field inhomogeneity. For nuclear sites which are randomly orientated with respect to the magnetic field the function $p(\xi)$ is given by[1]

$$p(\xi) = \sqrt{\frac{1}{2\xi+1}}. \qquad [12]$$

Computer simulations are limited by the assumptions that have to be made in calculating the spectrum. For instance residual interactions other than those of interest may affect the shape of the spectrum. If the spectrum consists of more than one powder component then difficulties may arise in assigning parameters to the individual components. The amount of time needed to compute such a multicomponent spectrum will also be increased considerably. To produce faithful simulations it is also important to ensure that the experimental data is not affected by such factors as receiver ring-down, power roll off and other instrumental distortions.

<u>Moment Analysis</u>

Moment analysis gives information on the shape of the resonance curve and on the rate at which it falls down to zero in the wings. For powder spectra it also gives information on the residual anisotropy. The nth moment of a spectrum (M_n) with a lineshape $f(\omega)$ is defined as

$$M_n = \int_{-\infty}^{\infty} \omega^n f(\omega)\ d\omega \bigg/ \int_{-\infty}^{\infty} f(\omega)\ d\omega\ , \qquad [13]$$

where $\omega=0$ corresponds to the Larmor frequency ω_0. It can be seen that all odd moments of a symmetric spectrum such as a powder spectrum from a spin 1 nucleus will be zero. Therefore for such spectra only one half of the spectrum is considered in calculating the moments.

For spin 1 powder patterns, moments can be used to provide information on the quadrupole splittings. If the powder spectrum is a true powder pattern with linebroadening that is much smaller than the apparent quadrupole splitting then the first moment M_1 gives the mean quadrupolar splitting from the relationship[22]

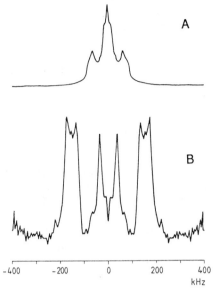

Figure 6 (A) 46.06 MHz ^{2}H NMR spectrum of hydrogen-exchange deuterated M13 coat protein as a solid powder at room temperature, using the quadrupolar echo technique. (B) Depaked spectrum derived from (A). Depaking was continued until no difference was observed between two successive iterations. The 12th iteration is shown at the left side and the 11th iteration at the right side.[25]

$$<\Delta v_Q> = \frac{3\sqrt{3}}{4\pi} M_1. \qquad\qquad [14]$$

The mean square quadrupolar splitting can be obtained from the second moment M_2 by

$$<(\Delta v_Q)^2> = \frac{5}{4\pi^2} M_2. \qquad\qquad [15]$$

Depaking

Depaking is an iterative procedure that is used to extract an oriented spectrum from a powder spectrum. This oriented spectrum corresponds to the resonance position when the symmetry axis of the tensor interaction is parallel to the external magnetic field. The procedure has been applied to spectra dominated by either the dipolar, quadrupolar or anisotropic chemical shift interactions.[23,24] Though the method assumes axial symmetry, scaling to 3

$\cos^2\theta$-1 and a random distribution of orientations the method can still be applied in a limited manner to partially oriented and axially asymmetric systems.

The method has been used to analyse powder spectra obtained from phospholipid systems. In the case of ^{31}P NMR, if there is a mixture of phospholipid phases such as lamellar and hexagonal H_{II} phase, then two superimposed powder patterns with anisotropies of opposite sign are obtained. The method of depaking can be used to resolve the signals from these two phases.

In ^{2}H NMR the method has been especially useful in resolving the signals from lipid acyl chains which have been perdeuterated and where many overlapping powder patterns are observed. The effect of depaking on a ^{2}H NMR spectrum of a ^{2}H-labeled protein is illustrated in Figure 6.[25] The broad component reflects an asymmetric tensor interaction, as seen by the splitting in the depaked spectra, whereas the narrow component indicates an axially symmetric powder pattern.

REFERENCES

1.   J. Seelig, Quart. Rev. Biophys., 1977, 10, 353.
2.   J. Seelig, Biochim. Biophys. Acta., 1978, 515, 105.
3.   R. G. Griffin, Meth. Enzymol., 1981, 72, 108.
4.   J. H. Davis, Biochim. Biophys. Acta., 1983, 737, 117.
5.   M. Mehring, "Principles of High Resolution in Solids", Springer-Verlag, Berlin, 1983.
6.   M. A. Hemminga and P. R. Cullis, J. Mag. Reson., 1982, 47, 307.
7.   A. Pines, M.G. Gibby and J.S. Waugh, Chem. Phys. Lett.,1972, 15, 373.
8.   R. G. Griffin, J. Am. Chem. Soc., 1976, 98, 851.
9.   P.R. Cullis and B. de Kruijff, Biochim. Biophys. Acta, 1979, 559, 339.
10.   G. W. Stockton, C. F. Polnaszek, A. P. Tulloch, F. Hasan and I. C. P. Smith, Biochemistry, 1976, 15, 954.
11.   E. E. Burnell, P. R. Cullis and B. de Kruijff, Biochim. Biophys. Acta, 1980, 603, 63.
12.   L. J. Burnett and B. H. Müller, J. Chem. Phys., 1971, 55, 5829.
13.   A. Abragam, "Principles of Nuclear Magnetism", Oxford University Press, 1961.
14.   M.A. Hemminga, P.A. de Jager, K.P. Datema and J. Breg, J. Magn. Reson., 1982, 50, 508.
15.   M.A. Hemminga and P.A. de Jager, J. Magn. Reson.,1983, 51, 339.

16. M.A. Hemminga, W.S. Veeman, H.W.M. Hilhorst and T.J. Schaafsma, Biophys. J. 1981, 35, 463.
17. M.A. Hemminga, K.P. Datema, P.W.B. ten Kortenaar, J. Krüse, G. Vriend, B.J.M. Verduin and P. Koole, in: Magnetic Resonance in Biology and Medicine (G. Govil, C.L. Khetrapal and A. Saran,eds.), Tata McGraw-Hill, New Delhi, 1985, pp. 53-76.
18. M.A. Hemminga, P.A. de Jager, J. Krüse and R.M.J.N. Lamerichs, J. Magn.Reson.,1987, 71, 446.
19. J. H. Davis, K. R. Jeffrey, M. Bloom, M. I. Valic and T. P. Higgs, Chem. Phys. Lett., 1976, 42, 390.
20. M. Rance and R. A. Byrd, J. Magn. Reson., 1983, 52, 221.
21. C. Schmidt, B. Blümich and H. W. Spiess, J. Magn. Reson., 1988, 79, 269.
22. J. H. Davis, Biophys. J., 1979, 27, 339.
23. M. Bloom, J. H. Davis and A. L. Mackay, Chem. Phys. Lett., 1981, 80, 198.
24. E. Sternin, M. Bloom and A. L. Mackay, J. Magn. Reson., 1983, 55, 274.
25. K.P. Datema, B.J.H. van Boxtel and M.A. Hemminga, J. Magn. Reson., 1988, 77, 372.

5
Dynamic Light Scattering by Partially-fluctuating Media

By P. N. Pusey

ROYAL SIGNALS AND RADAR ESTABLISHMENT, MALVERN, WORCESTERSHIRE WR14 3PS, U.K.

1 INTRODUCTION

A description of the principles underlying the technique of dynamic light scattering (DLS) was given in 1943[1]. However it was not until the invention of the laser in the early 1960's that the technique was developed as a practical proposition. Nowadays DLS is widely used in colloid and polymer science (and, to a lesser extent, in biology) as a powerful probe of the diffusive motions of particles, mainly in a liquid environment. The majority of applications still concerns systems in which the particles, given enough time, are able to diffuse throughout the sample; for this purpose the theory required for interpretation of the data is well developed. However there is growing interest in what may be called "partially-fluctuating media" such as polymer gels or colloidal glasses. The important characteristic of such media is that the scattering elements are not free to diffuse throughout the sample but are able only to execute limited random excursions about fixed average positions. The theoretical interpretation of DLS data obtained from such systems will form the main topic of this article.

In the next section, largely as background, we review briefly DLS by freely-diffusing systems. Section 3 then deals with partially-fluctuating media. A full report of this work will be published elsewhere[2]; here we will describe briefly the underlying principles and then simply quote and discuss the important results.

2 DLS BY FREELY-DIFFUSING SYSTEMS

References 3-7 give detailed descriptions of the subject matter covered briefly in this section.

In a DLS experiment laser light illuminates a sample containing, say, suspended colloidal particles or a solution of polymer molecules which scatter the light in all directions. For this application the important property of laser light is its coherence: phase relationships are maintained in the scattering process. Thus, at any instant, the far-field pattern of scattered light consists of a random diffraction or speckle pattern which is determined by the interference between the light fields scattered by different particles and therefore by their instantaneous positions and scattering amplitudes. As the particles move in Brownian motion, these individual fields change in phase (and, sometimes, in amplitude); in consequence the speckle pattern fluctuates from one random configuration to another. For sub-micron particles the characteristic time of these fluctuations is in the range of microseconds to milliseconds. A photon-counting photomultiplier tube, having sensitive area smaller than one speckle, is placed in the far field. The (standardised) output of this detector then consists of pulses of charge or voltage, each corresponding to the detection of a single photon, which are bunched up or spaced out in time in response to the randomly-fluctuating intensity at the photocathode. This digital signal is fed to a "photon correlator", a purpose-built computer, which constructs the time correlation function of the photodetections (see Eqn (4) below).

The amplitude at time t of the light field scattered by N particles can be written

$$E(K,t) = \sum_{j=1}^{N} a_j(K,t) \, \exp[i\underline{K}.\underline{r}_j(t)] \quad , \qquad (1)$$

where $a_j(K,t)$ is the amplitude of the light scattered by particle j and $\underline{r}_j(t)$ is the position of its centre-of-mass; $\underline{K}$ is the usual scattering vector, the difference between the propagation vectors of incident and scattered light, having magnitude

$$\underline{K} = \frac{4\pi}{\lambda} \sin \frac{\theta}{2} \, , \tag{2}$$

where λ is the wavelength in the medium of the light and θ the scattering angle. Fluctuations in the scattered light field $E(K,t)$ arise from two sources. The individual scattering amplitudes $a_j(K,t)$ may fluctuate, for example, as a non-spherical particle changes its orientation relative to $\underline{K}$ or as a flexible polymer undergoes conformational fluctuations. Secondly, translational Brownian motion of the particles leads to changing positions $\{\underline{r}_j(t)\}$, to fluctuating phase factors $\exp[i\underline{K}.\underline{r}_j(t)]$ and therefore to changes in the interference of the fields scattered by the different particles which determines the speckle pattern. The intensity corresponding to the field of Eqn (1) is

$$I(K,t) \equiv \left| E(K,t) \right|^2 . \tag{3}$$

The photon correlation function obtained from the correlator provides an estimate, for a range of delay time τ, of the normalised time correlation function of the scattered intensity

$$g_T^{(2)} (K,\tau) \equiv \frac{\langle I(K,0) \, I(K,\tau) \rangle_T}{\langle I(K) \rangle_T^2} \, , \tag{4}$$

where $\langle ... \rangle_T$ represents a time average. If the scatterers are free to diffuse over many wavelengths the phases $\{\underline{K}.\underline{r}_j(t)\}$ undergo fluctuations over many multiples of 2π. Then if, in addition, the number of scatterers N in the scattering volume is large, $E(K,t)$ is, by virtue of the central limit theorem, a zero-mean complex Gaussian random variable. Under these conditions the intensity correlation function can be written

$$g_T^{(2)} (K,\tau) = 1 + [f(K,\tau)]^2 \, , \tag{5}$$

where $f(K,\tau)$ is the normalised ensemble-averaged time correlation function of the field $E(K,t)$:

$$f(K,\tau) = \frac{\displaystyle\sum_{j=1}^{N}\sum_{k=1}^{N}\langle a_j(K,0)a_k(K,\tau)\exp\{i\underline{K}.[\underline{r}_j(0)-\underline{r}_k(\tau)]\}\rangle_E}{\displaystyle\sum_{j=1}^{N}\sum_{k=1}^{N}\langle a_j(K)a_k(K)\exp\{i\underline{K}.(\underline{r}_j-\underline{r}_k)\}\rangle_E} , \tag{6}$$

where $\langle...\rangle_E$ represents an ensemble average. This field correlation function has the limiting values

$$f(K,0) = 1 \tag{7}$$

and

$$f(K,\infty) = 0 , \tag{8}$$

where the first is by definition and the second follows from the assumption of freely-diffusing particles (large fluctuations of the phases $\{\underline{K}.\underline{r}_j(t)\}$). Use of Eqns (4), (5), (7) and (8) shows that the intensity correlation function $g_T^{(2)}(K,\tau)$ decays from a value

$$g_T^{(2)}(K,0) \equiv \frac{\langle I^2(K)\rangle_T}{\langle I(K)\rangle_T^2} = 2 \tag{9}$$

at $\tau = 0$ to

$$g_T^{(2)}(K,\infty) = \frac{\langle I(K)\rangle_T^2}{\langle I(K)\rangle_T^2} = 1 \tag{10}$$

as $\tau \to \infty$. The $\tau = 0$ limit of 2 is a consequence of the Gaussian property of the light field for freely-diffusing scatterers whereas the $\tau \to \infty$ limit follows from the lack of correlation between $I(K,0)$ and $I(K,\tau)$ when τ greatly exceeds the characteristic fluctuation time of the intensity.

In a DLS experiment, therefore, one can obtain an experimental estimate of $f(K,\tau)$ by measuring $g_T^{(2)}(K,\tau)$ and constructing the quantity $[g_T^{(2)}(K,\tau) - 1]^{\frac{1}{2}}$. Interpretation of the data then requires a theoretical model for $f(K,\tau)$. As can be seen from its definition, Eqn (6), $f(K,\tau)$ is, in general, a complicated quantity depending, for example, on the correlation between the orientation or configuration of one scatterer at time zero (through $a_j(K,0)$) with the position $\underline{r}_k(\tau)$ of another scatterer at time τ. Thus we look at simple cases.

The simplest useful application of DLS is to the study of dilute suspensions of identical spherical (or small compared to λ) particles. In this case $a_j(K,t)$ is independent of t and is the same for all particles; also there are no correlations between the positions or motions of different particles. Then $f(K,\tau)$ reflects only the translational motions of individual particles and it is straightfoward to show that

$$f(K,\tau) = \exp(-D_T K^2 \tau) \ , \qquad\qquad (11)$$

where D_T is the particles' translational diffusion constant. Data analysis consists simply of determining the slope $- D_T K^2$ of a plot of $\frac{1}{2} \ln [g_T^{(2)}(K,\tau) - 1]$ against τ. Since K can be calculated from λ and θ (Eqn (2)) one obtains an estimate of D_T, typically to an accuracy of 1 or 2% from a measurement lasting a couple of minutes. For spherical particles the radius R can be derived by use of the Stokes-Einstein relation

$$D_T = k_B T / 6\pi\eta R \qquad\qquad (12)$$

where $k_B T$ is the thermal energy and η the viscosity of the liquid. DLS is now probably the simplest and quickest method of sizing submicron spherical particles[8,9]. The friction coefficient of particles of arbitrary shape can be obtained from D_T and combined with sedimentation and density measurements to provide the molecular weight.

If the sample contains a distribution of particle sizes, $f(K,\tau)$ consists of a sum of exponentials, one corresponding to each size species present. In principle, data for $f(K,\tau)$ can be inverted to provide the particle size distribution; several methods exist for doing this[9]. However extremely accurate data are required and DLS is not, in general, a good method for resolving the details of narrow size distributions[10]. Nevertheless well-defined average properties, such as mean size and molecular weight, can be obtained.

For dilute suspensions of non-spherical or flexible particles which are comparable in size to λ, fluctuations in the scattered intensity arise not only from translational motions but also from rotational Brownian motion and conformational fluctuations (through the amplitude factors $\{a_j(K,t)\}$, as described above). Theoretical models have been developed which allow $f(K,\tau)$ to be analysed to provide information on rotational diffusion constants and the rates of conformational changes.

Considerable progress has been made in recent years in understanding DLS by interacting systems, such as concentrated suspensions of spherical colloids and concentrated polymer solutions, in which correlations between the positions and motions of different scatterers are important[7]. To review this area is beyond the scope of this article. It suffices to say that both collective properties, such as gradient diffusion, and single-particle properties, such as "tracer" diffusion, can be obtained from DLS studies of such systems. (In this connection note that the term "freely-diffusing" is used in this article to describe the ability of a scatterer, given enough time, to diffuse throughout the sample. It does not exclude the possibility that such diffusion may be strongly affected by interactions with other scatterers.)

3 DLS BY PARTIALLY-FLUCTUATING MEDIA

For freely-diffusing systems, considered in Sec. 2, the intensity observed at any point in the speckle pattern of scattered light undergoes with time the full range of fluctuations expected for a complex Gaussian field: the normalised intensity correlation function $g_T^{(2)}(K,\tau)$ decays from an initial value of $g_T^{(2)}(K,0) = 2$ to a long-time value $g_T^{(2)}(K,\infty) = 1$. Consider now the extreme case of a random arrangement of <u>totally stationary</u> scatterers. Again the scattered light will form a random speckle pattern consisting of bright spots and dark regions. However now, because the scatterers are not moving, the intensities at all

points in the pattern will be constant in time: the
time correlation functions of these intensities will
thus be independent of delay time τ, having normalised
values $g_T^{(2)}(K, \tau) = 1$. It follows that a partially-
fluctuating medium, in which the scattering elements
make limited excursions of spatial extent comparable to
or less than λ, will give rise to a partially-
fluctuating speckle pattern: the intensity at a point
in the pattern will undergo limited fluctuations about
a mean value which will itself vary from point to
point. As a consequence, the correlation function of
this intensity will have a zero-time value $g_T^{(2)}(K,0)$
$= \langle I^2(K) \rangle_T / \langle I(K) \rangle_T^2$ between one and two (but will again
decay to one as $\tau \to \infty$). Furthermore the actual form of
the measured intensity correlation function will depend
on the mean intensity of the speckle under study.

It appears that the proper theory for DLS by
partially-fluctuating media has not been given
previously. Recently we have worked out this theory[2]
and will summarise it briefly here. For simplicity we
assume that the amplitude of light scattered by the
j'th element is independent of time. Thus we write the
scattered field

$$E(K,t) = \sum_{j=1}^{N} a_j(K) \exp[i\underline{K}.\underline{r}_j(t)] \ . \tag{13}$$

To allow for the limited excursions of the scatterers
we take

$$\underline{r}_j(t) = \underline{r}_j + \underline{\Delta}_j(t) \ , \tag{14}$$

where $\underline{r}_j$ is the fixed mean position of scatterer j and
$\underline{\Delta}_j(t)$ is the fluctuation about this position.
Substitution of (14) in (13) gives

$$E(K,t) = \sum_{j} a_j(K) \exp[i\underline{K}.\underline{r}_j] \exp[i\underline{K}.\underline{\Delta}_j(t)] \tag{15}$$

which can be written

$$E(K,t) = \sum_{j} a_j(K)\exp[i\underline{K}.\underline{r}_j]\{\exp[i\underline{K}.\underline{\Delta}_j(t)] - <\exp[i\underline{K}.\underline{\Delta}_j>_T\}$$

$$+ \sum_{j} a_j(K)\exp[i\underline{K}.\underline{r}_j]<\exp[i\underline{K}.\underline{\Delta}_j]>_T \quad . \quad (16)$$

It can then be shown that the first term in (16) is a zero-mean complex Gaussian variable whereas the second term is clearly independent of time. Thus the total field E(K,t) is, by contrast with the case of a freely-diffusing system, no longer a Gaussian variable. (Note that, for a freely-diffusing system, $\underline{\Delta}_j(t)$ can vary over many wavelengths so that $<\exp(i\underline{K}.\underline{\Delta}_j)>_T$ is zero and (16) reduces to (1).)

Details of the derivation, which is not difficult, of the intensity correlation function corresponding to the field of Eqn (16) will be given elsewhere[2]. Here we simply quote the result:

$$g^{(2)}(K,\tau) = 1 + \frac{<I(K)>_E^2}{<I(K)>_T^2}\{[f(K,\tau)]^2 - [f(K,\infty)]^2\}$$

$$(17)$$

$$+ 2\frac{<I(K)>_E}{<I(K)>_T}\left\{1 - \frac{<I(K)>_E}{<I(K)>_T}\right\}\left\{f(K,\tau) - f(K,\infty)\right\} .$$

In Eqn (17) $<I(K)>_T$ is the (time) average intensity of the speckle being studied; $<I(K)>_E$ is the ensemble average intensity, ie the intensity averaged over many speckles; f(K, τ) is, as before, the ensemble-averaged field correlation function, Eqn (6) with $a_j(K,t) = a_j(K)$. Several general properties of the intensity correlation function given by (17) can be noted: (i) It can be shown that $g_T^{(2)}(K,0)$ is less than 2, corresponding to the restricted intensity fluctuations of one speckle; (ii) The form of $g_T^{(2)}(K,\tau)$ depends, through the ratio $\langle I(K)\rangle_E / \langle I(K)\rangle_T$, on the mean intensity $\langle I(K)\rangle_T$ of the speckle actually being observed; (iii) In the case of a partially-fluctuating medium $f(K,\tau)$ does not decay to zero as $\tau \to \infty$, but rather to a value $f(K,\infty)$ which is related to the magnitude of the "frozen-in" (non-decaying) fluctuations of the medium; (iv) For a freely-diffusing medium, the intensity of any one speckle undergoes, with time, the full range of fluctuations so that $\langle I(K)\rangle_T = \langle I(K)\rangle_E$; furthermore $f(K,\infty) = 0$. Then, as it must, (17) reduces to (5).

Equation (17) is a general result which relates $g_T^{(2)}(K,\tau)$, the quantity measured in a single DLS experiment, to $f(K,\tau)$, the quantity which must be modelled theoretically for the partially-fluctuating system under study. In this connection we point out that $f(K,\tau)$, also known as the intermediate scattering function, can be interpreted as the time correlation function of the spatial Fourier component of wavevector K of the refractive index or concentration fluctuations which cause the scattering (see Eqn (6)).

As an example we consider a situation for which a simple form for $f(K,\tau)$ can be derived. We assume that the scatterers are Brownian particles whose motions are independent and which are localised by harmonic forces centred on the (random) mean fixed positions $\{r_j\}$. This would be a simple model for dilute strongly-scattering "tracer" particles trapped in a weakly-scattering polymer gel[11]. For this model[12]

$$f(K,\tau) = \exp\{-K^2\langle\Delta^2\rangle[1-\exp(D_T K^2\tau/K^2\langle\Delta^2\rangle)]\} \ , \quad (18)$$

where $\langle\Delta^2\rangle^{\frac{1}{2}}$ is the root-mean-square displacement of a particle about its mean position and D_T is, as before, the translational diffusion constant of the particle when not constrained. We note the following properties of this function: (i) As $K^2\langle\Delta^2\rangle \to \infty$, corresponding to a freely-diffusing medium, $f(K,\tau) \to \exp(-D_T K^2\tau)$, as expected (see Eqn (11)); (ii) As $\tau \to 0$, $f(K,\tau) \to 1 - D_T K^2\tau$. Thus at short times (and over small distances) a particle diffuses as if it were not constrained; only at longer times is the spatial restriction imposed by the harmonic force felt; (iii) As $\tau \to \infty$, $f(K,\infty) = \exp(-K^2\langle\Delta^2\rangle)$, providing a direct measure of $\langle\Delta^2\rangle$ and hence the effect of the restriction.

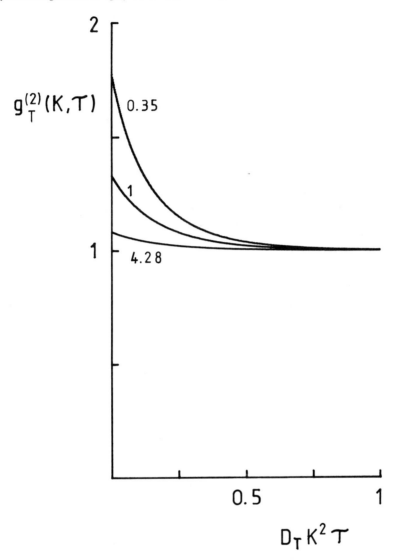

Figure 1 Time-averaged intensity correlation functions for a model partially-fluctuating medium. The curves are calculated from Eqns (17) and (18) with $K^2 \langle \Delta^2 \rangle = 0.2$; values of $\langle I(K) \rangle_T / \langle I(K) \rangle_E$ are indicated.

Figure 1 shows plots of the intensity correlation function $g_T^{(2)}(K,\tau)$ against $D_T K^2 \tau$, calculated from Eqns (17) and (18) for $K^2 < \Delta^2 > = 0.2$ and various values of $<I(K)>_T/<I(K)>_E$. The strong dependence of the measured $g_T^{(2)}$ on $<I(K)>_T$, the average intensity of the speckle under study, is evident; this could clearly complicate data analysis.

We observe, however, that if $<I(K)>_T = <I(K)>_E$, Eqn (17) takes on the simpler form

$$g_T^{(2)}(K,\tau) = 1 + [f(K,\tau)]^2 - [f(K,\infty)]^2 \qquad (19)$$

which differs from Eqn (5) only by the presence of the last term. This observation suggests the following general strategy for a DLS mesurement on a partially-fluctuating system:

(i) The sample is scanned at a steady rate through the laser beam so that a succession of many speckles illuminates the detector. Measurement of the intensity during this operation provides an estimate of the ensemble-averaged intensity $<I(K)>_E$.

(ii) The position of the sample (or the detector) is then adjusted carefully until the time-averaged intensity $<I(K)>_T$, seen by the detector, is equal to the ensemble-averaged value, determined in (i) above.

(iii) The intensity correlation function $g_T^{(2)}(K,\tau)$ is measured. The quantity $f(K,\tau)$, for comparison with theory, can then be obtained from

$$f(K,\tau) = [1 - g_T^{(2)}(K,0) + g_T^{(2)}(K,\tau)]^{\frac{1}{2}} \qquad (20)$$

which follows from rearrangement of Eqn (19).

Finally, we note that if, in ignorance of Eqn (17), one were to analyse data from a partially-fluctuating medium by the method usually used for a freely-diffusing system (ie by determining the slope of a plot of $\frac{1}{2}\ln[g_T^{(2)}(K,\tau) - 1]$ against τ, see Sec. 2) an incorrect value of the diffusion constant D_T would be obtained. Analysis by this method of the data shown in Figure 1 would give values of D_T too large by a factor of about three[2].

4 DISCUSSION

Although we have not emphasised it in this brief report, the difficulty in analysing DLS data from partially-fluctuating systems can be said to arise from their non-ergodicity – the non-equivalence of time and ensemble averages. DLS measures a time-averaged property whereas theory usually provides an ensemble average: for freely-diffusing media time and ensemble averages are equivalent; for non-ergodic media they are not. Thus observation of a single speckle in the pattern of light scattered by a partially-fluctuating medium provides only a partial average: the full range of fluctuations in the displacements $\{\Delta_j(t)\}$ is sampled but different sets of fixed positions $\{r_j\}$ are not. The treatment of Sec. 3 shows how the time-averaged intensity correlation function $g_T^{(2)}(K,\tau)$ (Eqn (4)), measured by DLS by a non-ergodic medium, is related (Eqn (17)) to the ensemble-averaged property $f(K,\tau)$ (Eqn (6)). Our full report[2] of this work will discuss non-ergodicity in more detail.

Our interest in non-ergodic media arose from DLS experiments on colloidal glasses[13], very concentrated suspensions of spherical colloids trapped in an amorphous structure and able only to move locally. There we circumvented the problems of non-ergodicity by performing a tedious "brute-force" ensemble average: measurements of intensity correlation functions made on many different speckles were summed. The new analysis described in Sec. 3 should allow much quicker and easier collection of data from these systems.

In recent years there have been several studies of polymer gels by DLS. It remains to determine the extent to which the conclusions drawn from these experiments are affected by incorrect analysis of the data. Recently Nishio _et al_.[11] described a study by DLS of the motions of tracer particles in polyacrylamide gels to which the simple model outlined in Sec. 3 should be directly applicable.

ACKNOWLEDGEMENTS

My interest in partially-fluctuating systems arose during collaborative work[13] with W van Megen and I thank him for many valuable discussions. I thank E Jakeman for an important suggestion concerning the theoretical development outlined in Sec. 3 and S J Candau for useful discussions.

REFERENCES

1.   G. N. Ramachandran, Proc. Indian Acad. Sci., 1943,
     18, 190.
2.   P. N. Pusey and W. van Megen, to be published.
3.   B. Chu, "Laser Light Scattering", Academic, New
     York, 1974.
4.   H. Z. Cummins and E. R. Pike, Editors, "Photon
     Correlation and Light Beating Spectroscopy",
     Plenum, New York, 1974.
5.   B. J. Berne and R. Pecora, "Dynamic Light
     Scattering", Wiley, New York, 1976.
6.   H. Z. Cummins and E. R. Pike, Editors, "Photon
     Correlation Spectroscopy and Velocimetry", Plenum,
     New York, 1977.
7.   R. Pecora, Editor, "Dynamic Light Scattering",
     Plenum, New York, 1985.
8.   H. G. Barth, Editor, "Modern Methods of Particle
     Size Analysis", Wiley, New York, 1984.
9.   B. E. Dahneke, Editor, "Measurement of Suspended
     Particles by Quasi-Elastic light Scattering",
     Wiley, New York, 1983.
10. See, however, the method described by P. N. Pusey
     and W. van Megen, J. Chem. Phys., 1984, 80, 3513.
11. I. Nishio, J. C. Reina and R. Bansil, Phys. Rev.
     Lett., 1987, 59, 684.
12. F. D. Carlson and A. B. Fraser, J. Mol. Biol.,
     1974, 89, 273.
13. P. N. Pusey and W. van Megen, Phys. Rev. Lett.,
     1987, 59, 2083.

6

Analysis of Macromolecular Structures in Solution by Electrooptical Procedures

By D. Porschke and J. Antosiewicz

MAX PLANCK INSTITUT FÜR BIOPHYSIKALISCHE CHEMIE, 34 GÖTTINGEN, F.R.G.

1. INTRODUCTION

Most of our knowledge on the structure of biological macromolecules comes from x-ray analysis of crystallized materials.[1,2] In spite of the great success of this procedure, there has always been a strong demand for an independent approach, which could be used to analyze macromolecular structures *in solution*. There are two major reasons for this demand. First of all, structures found in crystals need not be equivalent to structures in solution because of crystal packing interactions. Second and more important is the fact that the rigid and almost immobile structures found in crystals are not useful for most biological functions, which require flexible structures. Since none of the available methods provides all the requested informations on macromolecular structures in solution, a combination of methods should be used. For example, NMR analysis[3] is very useful to obtain information on the local structure. However, due to the limited accuracy, extrapolations to the long range structure are not easily possible. Furthermore, this method is still limited to relative small molecules. An approach, which is particularly well adapted to study macromolecules from the opposite viewpoint of the long range structure, is given by electrooptical procedures. In the present contribution, the potentials of these procedures are described and some examples are presented.

103

2. EXPERIMENTAL PROCEDURE

The experimental setup required for electrooptical measurements can be very simple, as shown in Fig. 1. The main components are an electric pulse generator and a spectrophotometric detection unit.[4,5] In dichroism experiments, the absorbance of polarized light is measured under electric field pulses. If the sample molecules are associated with a sufficiently high electric anisotropy, these molecules are aligned in the presence of the electric field. The alignment is detected by absorbance changes, provided that the molecules are also associated with an optical anisotropy. The conditions of sufficiently high electric and optical anisotropies are fulfilled for the majority of biological macromolecules; particularly high anisotropies giving rise to large electrooptical signals have been found for nucleic acids. Informations on the molecules under investigation can be obtained both from stationary and transient signals: the former ones are used to evaluate 'intrinsic' optical anisotropies and dipole moments or polarizabilities; the latter ones provide detailed information on the dimensions, the shape and eventually also the flexibility of the molecules.

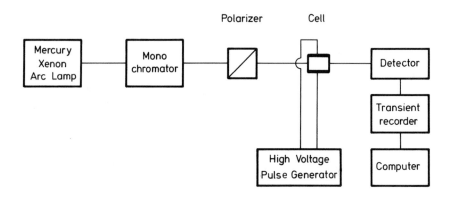

Figure 1 Scheme of an instrument for electrooptical measurements.

3. ANALYSIS OF ELECTRIC PARAMETERS

Usually the electric anisotropy of simple molecules is either due to a permanent or to an induced asymmetry of the charge distribution. Since all molecules are polarizable

– at least to some extent, molecules with a permanent dipole moment also have a contribution from an induced moment, which is, however, not always detectable. For biopolymers the nature of dipole moments is often much more complex. This is due to the fact, that biopolymers usually bear a large number of charged residues, which attract counterions. The electrostatic interactions between the various charges can be extremely complex and in most cases are hardly tractable – even in the absence of external electric fields. Thus, the theories on the electric properties of macromolecules have been developed for limit cases only (e.g. for low electric field strengths) and the experimental data usually have to be described by relatively simple phenomenological models.

For these reasons it is often very difficult to assign the nature of dipole moments of macromolecules. A well known example is the case of DNA double helices, which appear to be associated with permanent dipole moments, although double helices are symmetric with respect to their charge distribution.[6,7] More information on this phenomenon has been obtained recently by direct measurements of the polarization dynamics[8] in the ns-time range. A detailed analysis of dichroism rise curves observed for double helices with 76 and 95 bp revealed the existence of two separate relaxation processes. For this range of chain lengths much below the persistence length, double helices can be described as rigid rods reasonably well and their rotational diffusion observed in dichroism decay curves can be represented by single exponentials to a very good approximation. Obviously the difference between the rise and the decay curves must be due to the fact that the dichroism rise requires a polarization process, which is without concern for the decay curves. It should be expected that the rise curves can be described as a convolution product of a polarization and an orientation process. Indeed, all observed rise curves are consistent with this expectation.

The relaxation process reflecting polarization can now be exploited to learn more about the mechanism of the polarization process.[8] The dependence on the ion concentration demonstrates that polarization affects the bimolecular step of ion binding. Furthermore, the field strength dependence of the rate parameters shows that electric fields induce dissociation of ions. Finally, the field strength dependence of the dissociation rate constant is consistent with a reaction, which is driven by an increase of the DNA dipole moment. The dipole moment required for this reaction fits reasonably well to the dipole moment derived independently from the quantitative analysis of the stationary dichroism. All these conclusions are valid for the range of relatively high electric field strengths, where the DNA appears to be associated with a permanent dipole moment. At lower electric field strengths and/or low DNA chain lengths the polarization time constant was beyond the limit of time resolution of the presently available instrument. It is possible that in this range the ion polarization is essentially due to 'intramolecular' motion of counterions along the double helix.

Corresponding mechanisms may be expected for proteins. In this case a special variant of the ion polarization – the polarization of protons – may provide an important contribution. However, the proteins which have been investigated at high time resolution up to now, do not reveal any contributions from ion- or proton-polarization. Both the stationary dichroism and the dichroism rise curves indicate the existence of 'true' permanent dipole moments of *lac*[9] and *tet*[10] repressors. The dipole moments found for these proteins are surprisingly high – 1200 and 1050 debye units for *lac* and *tet* repressor, respectively. Obviously, these large dipole moments must have a considerable impact on the interactions of the repressors. Some other proteins are under investigation and also seem to be associated with large permanent dipole moments.

4. BENDING OF DNA DOUBLE HELICES

As already mentioned in the previous section, short DNA double helices with chain lengths much below the persistence length may be regarded as rigid rods. In this limit case the dichroism decay can be described by single exponentials.[6,13] However, deviations from single exponential decay curves can be easily detected for DNA helices with chain lengths around the persistence length.[12] These deviations are not only dependent on the DNA chain length, but also on the electric field strength of the pulses used for the molecular alignment and on the ionic conditions of the solutions.

With respect to the ionic conditions there are two limit cases. At low ion concentrations the dipole induced by the electric field does not only lead to alignment of the double helices, but also to stretching of bent molecules. Due to the strong chain length dependence of molecular motions, stretched helices return more quickly to their bending equilibrium upon pulse termination than aligned molecules return to their random distribution. Thus, bending of stretched helices can be detected as an individual relaxation process, which is clearly separated on time scale from the overall distribution process. Bending amplitudes and time constants of double helices have been identified as a function of the chain length. A DNA fragment of 179 base pairs, for example, shows a bending amplitude[12] corresponding to 30% of the total dichroism amplitude at field strengths of 70 kV/cm and a bending time constant of 180 ns. Recently, a simple orientation function for weakly bent rods has been developed,[13] which explains both chain length and field strength dependence of bending amplitudes. Some deviations observed for special ionic conditions are probably due to the existence of stably curved DNA sequences.

Another type of dichroism decay curves has been observed at high concentrations of monovalent ions. Under these conditions electric fields of relatively high amplitude and/or pulse length induce decay curves with an inversion of the dichroism, suggesting a component with a positive dichroism.[14] These results have been explained by bead model simulations on bent DNA double helices. For these calculations it has been assumed that the double helices are bent smoothly into circular arcs. Obviously, the DNA looses symmetry upon bending, which leads to a permanent dipole moment associated with bent DNA. Since DNA molecules bear a net negative electric charge, dipole moments with a physical significance have to be calculated with respect to the center of diffusion, which has been defined by bead model calculations. According to these calculations the permanent dipole moments associated with bent DNA can be very large[14] and amount to about 3100 debye units for a 179 bp fragment, which is bent to a half circle. When these dipole moments are considered in simulations of dichroism decay curves, the unusual dichroism effect described above is readily generated.

It remains to be explained, why the dichroism response is so much dependent on the salt concentration. Independent measurements show that the DNA polarizability strongly decreases with increasing monovalent salt concentration, which corresponds to a decrease of the dipole moment induced under electric field pulses at the ends of the DNA molecules. According to polyelectrolyte theory, the phosphate charges remaining after ion condensation are virtually independent of the salt concentration and, thus, the permanent dipole associated with bent DNA is also almost independent of the salt concentration. For this reason the influence of the permanent dipole moment relative to that of the induced dipole moment increases with increasing salt concentration, which appears to be the main cause for the induction of the unusual dichroism decay curves at high monovalent salt concentrations. Another important factor for the induction of the unusual dichroism curves seems to be field induced bending of double helices, which is driven by the increase of the dipole moment upon bending. This example shows that a quantitative interpretation of dichroism decay curves requires detailed knowledge of the electric parameters for the molecules under investigation.

5. DNA PROTEIN COMPLEXES

From measurements of the electrophoretic mobility in gels it has been concluded that binding of the cyclic AMP receptor protein to its specific DNA target induces a special conformation change of the DNA corresponding to bending of the double helix.[15] Since the electrophoretic mobility of macromolecular complexes in gels may be influenced by many different factors, it is virtually impossible to interpret these mobilities quantitatively. For a quantitative analysis of the special complex, electrooptical procedures proved to be very useful. As mentioned already above, dichroism decay time constants, which reflect rotational diffusion, are very sensitive to molecular dimensions. The rotational diffusion time constants of any complex are dominated by the component which is most extended in space. For protein-DNA complexes this component usually is the DNA double helix. The rotational diffusion can be measured for the free DNA and the DNA associated with protein. In most cases a simple comparison of these data is already sufficient to detect major changes of the DNA conformation. If the contribution from the protein component to the rotational diffusion coefficient cannot be neglected, some independent information on its size and shape has to be used – for example from electrooptical data measured for the protein alone or from x-ray analysis of its crystal structure. In the case of the cAMP receptor protein, the data analysis was facilitated by the fact, that the conformation change of the protein-DNA complex requires addition of a cofactor of relatively low molecular weight: cAMP. Thus, the rotational diffusion data for the complex before and after addition of cAMP may be compared directly, without necessity to introduce corrections for the cAMP component.

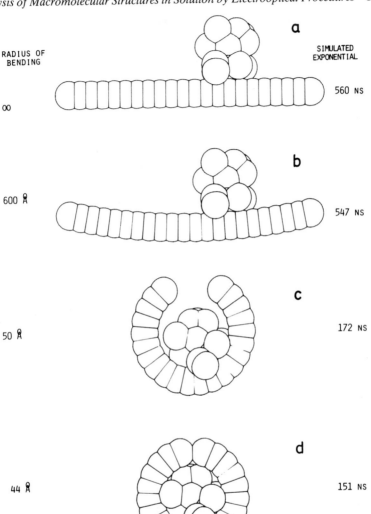

RADIUS OF
BENDING

SIMULATED
EXPONENTIAL

a

560 NS

00

b

600 Å

547 NS

c

50 Å

172 NS

d

44 Å

151 NS

Figure 2 Bead model of cyclic AMP receptor associated with straight 80 bp DNA (a). The same complex at different degrees of smooth bending of the DNA is shown in b (600 Å), c (50 Å) and d (44 Å) together with the simulated dichroism decay time constant. The experimental value is 130 ns (from Antosiewicz and Porschke[18]).

The electrooptical data[16] obtained for cAMP receptor-DNA complexes under various salt conditions clearly demonstrate that addition of cAMP induces a strong reduction of the complex dimensions. These data have been analyzed quantitatively by bead model simulations[17,18] using the known crystal structure of the receptor protein.[19] An example is shown in Fig. 2 for a complex formed from a 80 bp DNA fragment and the cAMP receptor at relatively high salt concentration. The complex has been simulated by overlapping beads of 14.2 Å radius and the protein component has been attached with its helix-turn-helix motif at the center of the DNA palindrom.[19] Since the dichroism decay constant simulated for a complex with straight DNA is much too large, the DNA double helix must be bent. The most convenient mode for calculations is smooth bending into circular arcs. As shown in Fig. 2 the double helix has to be bent as much as possible corresponding to wrapping of the DNA around the protein, in order to simulate the experimental time constant.

Another example of a bead model is given in Fig. 3 for the complex formed from a 203 bp DNA fragment with two specific binding sites and two protein dimers.[19] In this case the model has been constructed under the assumption that the two

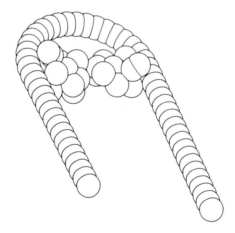

Figure 3 Bead model of the complex formed from a DNA fragment of 203 bp with two specific sites and two cAMP receptor protein dimers. The simulated time constant for the dichroism decay is 1.43 μs and the corresponding experimental value is 1.48 μs (from Antosiewicz and Porschke[18]).

protein dimers come into contact with each other and form a tetramer with two DNA binding sites. If the tetramer has tetrahedral symmetry, its binding sites are automatically in the proper orientation for association with two strands of DNA running in opposite direction. Although the DNA binding is somewhat distorted according to this model, the tetrahedral symmetry of the protein component appears to be attractive. Furthermore, the model suggests potential strong binding sites for polymerase. However, it should be emphasized that the bead model shown in Fig. 3 remains hypothetical. The protein and the DNA may also be organized in a different structure – for example the double helix may be wrapped around a relatively compact protein tetramer with the DNA ends protruding in opposite directions. Although the number of possible structures may appear to be very large at a first glance, this number is restricted to a large extent, because the external dimensions of the complex must be similar to those of the bead model shown in Fig. 3.

Some comments on the bead models of the cAMP receptor DNA complexes should be added. The binding geometries of the models shown in Figs. 2 and 3 are different and thus the models are not consistent with the expectation that binding geometries are equivalent. It should be emphasized, however, that the central portion of the binding domains can be very similar according to the proposed models and the difference may be mainly restricted to the 'outer' contacts between protein and DNA. Furthermore, the equivalence of structures can be quite misleading, if considered as a dogma. The static models derived from x-ray analysis of crystals are not necessarily valid in solution. In the case of the cAMP receptor – DNA complexes the electrooptical data clearly reveal that the structures in solution are strongly dependent on the environmental conditions.

According to the model given in Fig. 2, the double helix is wrapped around the protein at a remarkably low bending radius. If the corresponding free energy of bending is estimated as a standard mechanical energy from the persistence length,[12] the bias against this structure appears to be prohibitively high. However, according to current models of chromatin, the DNA double helices are wrapped around the nucleosomes at a very similar bending radius.[20] Apparently the harmonic approximation of bending energies is misleading at very high degrees of bending. Probably kinks introduced in the double helix at given intervals serve to reduce the energy required for the transfer of DNA to very compact forms.

For comparison, a bead model based on electrooptical data is shown in Fig. 4 for the complex formed by *tet* repressor protein and *tet* operator DNA. In this case all the species including the protein have been characterized by electrooptical measurements.[10] It is very likely that the DNA binding site of the *tet* repressor protein is folded as a helix-turn helix motif and thus should be similar to that of the cAMP receptor protein. Nevertheless binding of the *tet* repressor to *tet* operator DNA does not induce any major degree of bending of the double helix.

6. SUMMARY

Electrooptical measurements can be very useful for the characterization of macromolecules and of their complexes in solutions. First of all, electrooptical data can be used to characterize the electric properties, which are often very complex for natural macromolecules – mainly because of their polyelectrolyte nature. In the case of relatively short DNA double helices it has been possible to identify the polarization mechanism by a detailed analysis of the dichroism rise curves at ns-time resolution. Some proteins like *lac* and *tet* repressor prove to be associated with unusually large permanent dipole moments of more than 1000 Debye units. It is very likely that these dipole moments are very important for molecular interactions.

Detailed informations on the long range structure of macromolecules and of their complexes are accessible by a quantitative analysis of dichroism decay curves. This approach proves to be particularly successful in the investigation of protein-nucleic acid complexes, where the optical properties and to a large degree also the electric properties are dominated by the nucleic acid component. As shown for the cases of cAMP receptor-promotor complex and the *tet* repressor-operator complex, bead model simulations of dichroism decay curves can be used to derive detailed models of the long range structure.

Furthermore electrooptical data provide information on molecular flexibilities. In the case of DNA double helices it has been possible to determine bending amplitudes and time constants. Finally electrooptical procedures can also be used to characterize reactions and conformation changes induced by electric field. The very large number of different molecular responses, which may be induced by electric field pulses, can be quite dangerous, if the effects are not clearly analyzed and separated.

BEAD MODEL   EXPERIMENT

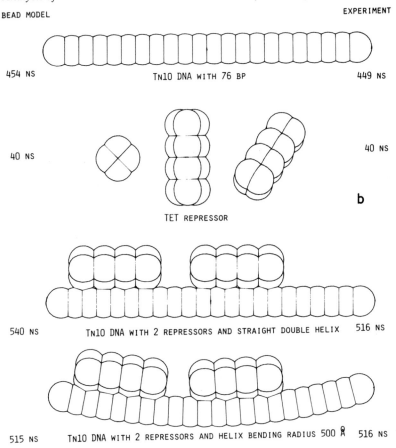

454 NS   Tn10 DNA WITH 76 BP   449 NS

40 NS   40 NS

b

TET REPRESSOR

540 NS   Tn10 DNA WITH 2 REPRESSORS AND STRAIGHT DOUBLE HELIX   516 NS

515 NS   Tn10 DNA WITH 2 REPRESSORS AND HELIX BENDING RADIUS 500 Å   516 NS

50 Å

<u>Figure 4</u> Bead models of Tn10 DNA with 76 bp, the *tet* repressor protein in 3 diffe-
rent views, the complex formed from two repressors with straight Tn10 DNA and the
same complex after bending of the DNA at a bending radius of 500 Å. The dichroism
decay time constants for the bead model and the corresponding experimental values
are on the left and right side, respectively (Porschke <u>et al.</u>[10]). Since the rotation
time constant of the complex is mainly determined by the long dimension of the
DNA, we cannot rule out the possibility that the repressor dimers are located on
opposite sides of the double helix.

However, it is possible to separate these effects by carefully designed experiments and then electrooptical data combined with thorough theoretical calculations provide detailed information on macromolecules in solution, which are often hardly accessible by other approaches.

REFERENCES

1. W. Saenger, 'Principles of Nucleic Acid Structure', Springer, Berlin, 1984.

2. G.E. Schulz and R.H. Schirmer, 'Principles of Protein Structure', Springer, 1979.

3. D.E. Wemmer and B.R. Reid, Ann. Rev. Phys. Chem. 1985, 36, 105.

4. E. Fredericq and C. Houssier, Electric dichroism and electric birefringence, Clarendon, Oxford, 1973.

5. C.T. O'Konski, Ed., 'Molecular Electro-optics', Dekker, New York, 1976.

6. M. Hogan, N. Dattagupta and D.M. Crothers, Proc. Natl. Acad. Sci. USA 1978, 75, 195.

7. S. Diekmann, W. Hillen, M. Jung, R.D. Wells and D. Porschke, Biophys. Chem. 1982, 15, 157.

8. D. Porschke, Biophys. Chem. 1985, 22, 237.

9. D. Porschke, Biophys. Chem. 1987, 28, 137.

10. D. Porschke, K. Tovar and J. Antosiewicz, Biochemistry 1988, 27, 4647.

11. S. Diekmann, W. Hillen, B. Morgeneyer, R.D. Wells and D. Porschke, Biophys. Chem. 1982, 15, 263.

12. D. Porschke, J. Biomol. Structure & Dynamics 1986, 4, 373.

13. D. Porschke, Biopolymers, in press.

14. J. Antosiewicz and D. Porschke, Biophys. Chem., in press.

15. H.M. Wu and D.M. Crothers, Nature 1984, 308, 509.

16. D. Porschke, W. Hillen and M. Takahashi, EMBO J. 1984, 3, 2873.

17. J. Garcia de la Torre and V.A. Bloomfield, Quart. Rev. Biophys. 1981, 14, 81.

18. J. Antosiewicz and D. Porschke, J. Biomol. Structure & Dynamics 1988, 5, 819.

19. D.M. McKay and T.A. Steitz, Nature 1981, 290, 744.

20. T.J. Richmond, J.T. Finch, B. Rushton, D. Rhodes and A. Klug, Nature 1984, 311, 532.

7

Fluorescence Studies of Protein-subunit Interactions

By M. K. Han,[1] J. R. Knutson,[2] and L. Brand[3]

[1]SECTION ON PROTEIN CHEMISTRY, LABORATORY OF BIOCHEMISTRY, NATIONAL HEART, LUNG AND BLOOD INSTITUTE, NATIONAL INSTITUTES OF HEALTH, BETHESDA, MD. 20892, U.S.A.

[2]LABORATORY OF TECHNICAL DEVELOPMENT, NATIONAL INSTITUTES OF HEALTH, BUILDING 10, ROOM 5D-10, BETHESDA, MD. 20892, U.S.A.

[3]DEPARTMENT OF BIOLOGY, THE JOHNS HOPKINS UNIVERSITY, BALTIMORE, MD. 21218, U.S.A.

I INTRODUCTION

Fluorescence:

Interaction of light with matter can be characterized in terms of energy, probability and direction. Conformational changes of macromolecules can influence each of these attributes of either intrinsic (typically tryptophan) or extrinsic fluorescence probes. Measurements of fluorescence excitation and emission spectra provide information about the energy of electronic transitions. Quantum yields (the number of quanta emitted/the number of quanta absorbed) or decay times give information about the probability of fluorescence emission versus other processes that an excited molecule may undergo. Fluorescence anisotropy gives information about directions between excitation and emission transition dipoles and the decay of the emission anisotropy reflects the loss of photoselection-induced anisotropy with time and thus provides detailed information about the rotational behavior of the chromophore.

Fluorescence can be measured either by steady-state methods or with time-resolved techniques. Steady-state measurements provide an intensity weighted average of the underlying decay processes. Thus these signals are dominated by the component emitting the most light and not by the most populated decay component. Nanosecond time-resolved techniques have the advantage that the steady-state information is dissected into the individual decay components. Time-resolved measurements are a particular asset for unraveling more detailed aspects of the rotational diffusion of protein molecules.

Fluorescence measurements are influenced not only by microheterogeneity around the ground-state environment of the fluorophore but also by the kinetics of any excited-state interactions that take place on the nanosecond time scale. Numerous excited-state processes such as resonance energy transfer, excimer formation, proton transfer and Brownian rotational motion are well understood and may be used to advantage to obtain information about the conformation of macromolecules[1,2]. There have been significant improvements in both instrumental methods and techniques for data analysis. Both pulse and phase/modulation instrumentation allow data to be obtained well into the picosecond time domain and the quality of the data is such that complex decay mechanisms can be evaluated[3].

A large number of numerical techniques including nonlinear least squares[4,5], the method of moments[6], Laplace transforms[7-10], and modulating functions[11] have been used for analysis of both fluorescence intensity and emission anisotropy decay data. Of greater interest is the fact that it is now appreciated that the simultaneous analysis of several fluorescence decay experiments (global analysis)[12,13] can increase the ability to discriminate between several proposed mechanisms and to recover the quantitative parameters that define the system under study.

When the time-resolved fluorescence of a protein is examined one is often faced with a bewildering number of decay parameters that can be recovered. What approach is feasible to understand the origin of complex decay and to relate the decay to the structure and function of a protein ?

A valuable approach has been that of association and overdetermination. For proteins that contain more than one tryptophan residue, fluorescence decay data can be obtained as a function of emission wavelength. Decay associated spectra (DAS)[14,15] can be generated from this data matrix. In favorable cases decay parameters can be assigned to specific tryptophan residues with unique emission spectra. In some cases one tryptophan residue may be more susceptible to quenching by an external agent such as iodide or acrylamide than other tryptophan residues. In this case fluorescence emission or excitation spectra are obtained as a function of quencher concentration and quenching decay associated spectra (QDAS)[16] can be obtained.

This approach has been extended to fluorescence emission anisotropy decay measurements to unravel possible rotational heterogeneity in biological systems. The approach is quite similar to that worked out for DAS. In this case, anisotropy decay-associated spectra (ADAS) are obtained and spectra associated with unique rotating species are resolved[17,18].

This general concept of association and overdetermination has been extended to include the use of a second time axis i.e. reaction time as a global overdetermination tool. The procedure used for this new application of global analysis are, once again, quite similar to those used in DAS, SAS, and ADAS. Fluorescence decay curves obtained at different times during the course of a chemical reaction are presumed to contain the same (but unknown) fluorescence decay time components. Their proportions (relative amplitudes) may change with reaction time. The amplitude values are unrestricted. In the case of proteins with multiple tryptophans, site-specific information may be obtained which is often difficult to achieve in a system of reacting biological macromolecules by other methods.

For excited-state reaction systems such as excimers, exciplexes and excited-state proton transfer, individual decay times are not associated with specific emitting species but are characteristic of the system as a whole. To discuss these situations, it is convenient to introduce a new term, species-associated spectra (SAS)[19-21]. The data matrix made up of fluorescence intensity versus time versus emission wavelength can be expressed in terms of time-resolved emission spectra (TRES), DAS, or SAS.

Fluorescence studies of proteins can be done using intrinsic fluorophores such as tryptophan or tyrosine or using fluorescent coenzymes such as NADH. The advantage of this approach is that the study is done with the native protein or protein-substrate complex. The disadvantage is that the intrinsic probes may not have the spectroscopic characteristics most desirable for the particular problem under study.

An alternative approach is to use extrinsic probes tailor made for each application and covalently or noncovalently attached to the protein at specific and unique sites. Among the aromatic residues of proteins, tryptophan is that most frequently used for fluorescence studies due to its relatively high quantum yield. However, some proteins lack tryptophan residues. In addition, the mean decay time of tryptophan is usually 5-6 ns, while the rotational correlation time of a typical protein with MW=100,000 is close to 150 ns. Therefore, the results obtained from emission anisotropy measurements with tryptophan have a significant uncertainty. A long lived, covalently bound extrinsic probe is desirable for studies of hydrodynamic properties of such molecules.

As examples of the overdetermination approach to the study of proteins we will describe recent work on two proteins. The first is Enzyme I of the bacterial phosphotransferase system. Subunit association of this protein appears to have an important role in the control of sugar translocation across the bacterial membrane[22]. A second example that will be discussed is horse liver alcohol dehydrogenase. This is a well studied enzyme in which Zn atoms have a role in maintaining a dimeric structure[23].

II. FLUORESCENCE STUDIES OF ENZYME I OF THE PTS.

A. Fluorescence anisotropy studies of Enzyme I.

The phosphoenolpyruvate: glycose phosphotransferase system (PTS) was discovered by Roseman and his coworkers in the early 1960's[24-26]. The main function of the PTS is to mediate the concomitant phosphorylation and translocation of sugar substrate across the bacterial membrane. The PTS is unique in that its phosphate donor is phosphoenolpyruvate and its sugars are primarily of the D-gluco and D-manno configuration. The PTS is responsible for chemotaxis toward PTS sugars in enteric bacteria, and plays an important role in the regulation of adenylate cyclase and in the synthesis of inducible catabolic enzymes required for utilization of certain non-PTS sugars and some non-PTS transport systems[27-29].

The overall phosphorylation reaction of the PTS involves a series of phosphoryl transfer steps and requires several protein components: Enzyme I and HPr are soluble, non-sugar-specific proteins. Another protein, Enzyme II is usually found to consist of two sugar-specific proteins constituting (with a lipid) an Enzyme II complex (III/II-B or II-A/II-B).

The first protein component of the sequence, Enzyme I is a dimer of identical subunits at room temperature. It accepts one phosphoryl group per

subunit from PEP in the presence of Mg^{2+} [30-32]. The
phosphoryl group is linked to the N-3 position of the
imidazole ring in a histidine moiety[33] and is donated to
HPr. The phosphoryl group is then ultimately transferred to sugars.

Neyroz _et al._[34] characterized the temperature-
dependent association of Enzyme I with emission
anisotropy methods. The nanosecond time-resolved
fluorescence of intrinsic tryptophan fluorescence was
used in these studies. Fig. 1 shows the mean rotational
correlation time (left side) and the changes in concentration of monomer and dimer species (right side)
of Enzyme I as a function of temperature. It was not
possible to resolve more than a single rotational time
for each decay experiment with a single curve analysis
and consequently, only the mean rotational correlation
time could be recovered as indicated in the left
figure. The resolution of two rotational correlation
times was achieved with the aid of global analysis
methods. The simultaneous analysis of _all_ the emis-
ssion anisotropy decay data was based on a model
describing a monomer/dimer mixture with a linkage of
rotational correlation times. The amplitude, β,
represents the percent of dimer species in solution as
a function of temperature. An error analysis indicated a large uncertainty in the estimation of the
rotational correlation times. This is due to the short
mean lifetime of tryptophan fluorescence in Enzyme I
(~5.9 ns) compared to the rotational correlation time
of the monomer which is well over 80 ns. While this
experiment was done with the native protein, the weak
point of the data is related to the short intensity
decay time of the fluorophore. A covalently bound
extrinsic probe with a longer decay time is desirable
for studies of the hydrodynamic properties of Enzyme I.
This approach will also be required for studies of most
medium or large proteins.

B. Site-specific labeling of Enzyme I.

Why is it important to label proteins on specific
sites ? Usually fluorescence studies with a fluorescent protein conjugate are aimed at relating structure
to function. In the case of Enzyme I, it is of interest
to determine whether the dimer is the active species
and whether dissociation into monomers always involves
loss of PTS activity. It is clearly essential to be
assured that the protein species being used to measure
the monomer/dimer association is the same as that
reporting the activity. This certainty can only be
achieved if site-specific labeling is assured. There
is a possible disadvantage often cited for site spe-

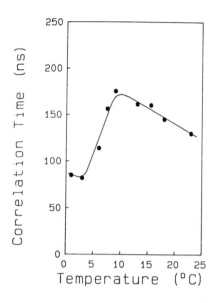
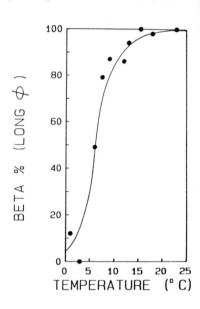

FIGURE 1: Mean rotational correlation time (left) and association (right) of <u>E. coli</u> Enzyme I as a function of temperature. Left, Experiments were performed with 1 mg/ml of <u>E. coli</u> Enzyme I in 75 mM potassium phosphate, pH 6.5, 1 mM EDTA, 0.5 mM dithiothreitol, and 5 mM $MgCl_2$. Each data set was analyzed in terms of a single rotational correlation time by the sum and difference procedure. The χ^2 for analysis of the difference curve ranged from 0.97 to 1.1. Right, anisotropy decay data was analyzed as $r(t) = \beta_1 e^{-t/\phi_1} + \beta_2 e^{-t/\phi_2}$. The $\%\beta_2$ (Y axis) represents the relative concentration of dimer Enzyme I at each temperature. The parameters describing the fluorescence decay and the decay of the emission anisotropy were obtained by global methods. The 67% joint confidence intervals of the two rotational correlation times obtained at 23°C were $\phi_1 = 21 < 43 < 77$ and $\phi_2 = 83 < 127 < 185$. This figure is taken from Neyroz <u>et al</u>. (1987) with permission from authors.

cific labeling. If the macromolecule under investigation is an anisotropic rotator i.e. if it is non spherical, the apparent rotational behavior can be influenced by the relation between the transition dipole of the probe and the rotating axes of the macromolecule. This difficulty is due to a "weighting" problem and is not a problem at all if the correct rotational decay can be revealed with the aid of global analysis[35].

Titration of Enzyme I with DTNB revealed four reactive -SH groups per monomer[36,37], in agreement with the predicted amino acid composition based on the enzyme's DNA sequence. The -SH residues appeared to be good candidates for the attachment of covalent labels.

The kinetics of the DTNB reaction with Enzyme I exhibit biphasic character, with pseudo-first order rate constants of 2.3×10^{-2}/s and 2.3×10^{-3}/s at pH 7.5, at room temperature. Fractional amplitudes associated with the rate constants were $25 \pm 5\%$ for the fast and $75 \pm 5\%$ for the slow rate.

The "slow" rate was influenced by HPr, Mg^{2+}, Mg^{2+} plus PEP, and also by temperature (at the temperature range where the monomer/dimer association occurs). Preincubation of the enzyme with 5 mM Mg^{2+} decreased the slow rate 2-3 fold. Moreover, addition of PEP in the presence of Mg^{2+} enhanced the effect an additional two-fold (compared to Mg^{2+} alone), while PEP alone showed virtually no effect. By contrast, preincubation with HPr alone (and pyruvate alone) resulted in an increase in the slow rate, rather than a decrease. The fractional ratio of the two rates remained at 1:3 under these conditions. These and related experiments indicated that one of the four -SH residues reacted more rapidly than the other three with -SH reagents and results in an active enzyme capable of forming dimer, whereas the modification of the slow -SH groups results in inactive and monomeric Enzyme. The confirmation of site-specific labeling of Enzyme I was provided by reverse-phase HPLC analysis of tryptic peptides and sequence analysis of fluorescent peptides[36,37]. These studies identified the fast reacting -SH group as the C-terminal cysteine. This residue was used as the site of specific labeling of Enzyme I with fluorescence probes.

C. Fluorescence studies of labeled Enzyme I.

Pyrene maleimide, a -SH group-specific reagent, was chosen as an extrinsic probe because it has a long decay time[38-41] and is thus useful for time-resolved

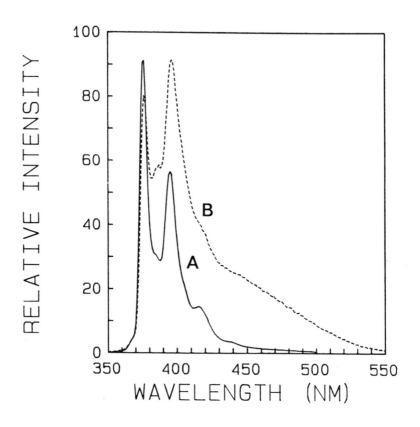

FIGURE 2 : Fluorescence emission spectra of pyrene
maleimide conjugated on the fast (A) and slow (B) -SH
groups of Enzyme I in 0.1M potassium phosphate buffer,
pH 7.5, containing 1 mM EDTA at 23°C. Excitation
wavelength was 340nm. The fluorescnce intnsity decreases upon ring opening. The peak (396nm) of B was
normalized to the peak (376nm) of A.

emission anisotropy studies as well as for simpler steady-state fluorescence anisotropy studies. A potential difficulty with the use of pyrene maleimide as a probe is that ring opening can occur after conjugation. It has been suggested that $\underline{S}$-[$\underline{N}$-(1-pyrene)-succinimido]-cysteine, a product of the reaction of the sulfhydryl group of cysteine with the olefinic double bond of the maleimide moiety of $\underline{N}$-(1-pyrene)maleimide, undergoes a slow cleavage of the succinimido ring by either hydrolysis or aminolysis. Aminolysis accompanies cyclization of the succinimido ring to form thiazine derivatives as a result of subsequent nucleophilic attack by the amino group on a carbonyl carbon[42,43]. The nature of these rearrangements often depends on the environment of the cysteine residue[44] (i.e., proximal lysine residues). The ring opening and rearrangements of pyrene maleimide lead to spectral changes and most likely to changes in decay time.

Figure 2 shows emission spectra of pyrene maleimide conjugated mainly to the fast- (A) and also mainly to the slow- reacting (B) -SH groups of Enzyme I at pH 7.5. When Enzyme I is labeled with less than one mole PM/monomer at pH 7.5, preferential labeling takes place at the C-terminal cysteine residue. The spectrum for this case (Fig. 2A) shows that the emission peak at 376nm is higher than the peak at 396 nm. In contrast, when slow -SH groups are labeled at pH 7.5 (with the fast -SH group protected by reaction with DTNB), the ratio of these two emission peak heights is quite different (Fig. 2B). These -SH residues are characterized not only by their decreased reactivity with sulfhydryl reagents but also by the more rapid rate of ring opening as indicated by significant changes in the emission spectra[36]. The pyrene maleimide conjugated at the "fast" -SH is reasonably stable for 7 days at pH 7.5 and even more stable at slightly more acidic pH (6.5). On the other hand, the imide ring on the slow - SH groups opened more rapidly, so that after 48 hour (pH 7.5), the derivative had lost about 50% of the initial emission. In addition at least two of the "slow" -SH groups (Fig. 2B) are located close enough to each other so that intramolecular excimer formation can take place when the ring opening occurs. We have not observed excimer formation when pyrene is conjugated only to the "fast" -SH residue.

The ring opening reaction is quite slow with the pyrene maleimide adduct at the C-terminal cysteine, and it was neglected in these studies. The monomer/dimer association of the labeled enzyme was characterized with fluorescence emission anisotropy using pyrene

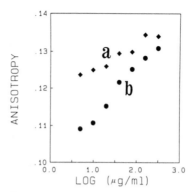

FIGURE 3: Experiments were performed with Enzyme I specifically labeled on the Cys-575 in 0.1 M potassium phosphate buffer, pH 7.5 containing 1 mM EDTA at 23°C. A: 5mM MgCl$_2$ was added and B: no MgCl$_2$ was added.

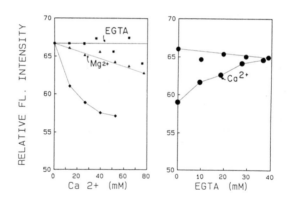

FIGURE 4: Experiments were performed with 0.1 mg/ml of Enzyme I in 0.1 M HEPES, pH 7.5, containing 1 mM EDTA at 23°C. Left, tryptophan fluorescence was measured with Enzyme I as a function of Ca^{2+} concentration in the absence of ligand (diamonds), in the presence of 20 mM MgCl$_2$ (triangles), and in the presence of 40 mM EGTA. Right, tryptophan fluorescence was measured with Enzyme I preincubated with and without 50 mM CaCl$_2$ (both samples included 40mM MgCl$_2$) as a function of EGTA concentration. Excitation and emission wavelengths were 295nm and 340nm, respectively.

fluorescence. As expected for a protein undergoing a monomer/dimer equilibrium, the emission anisotropy of fluorescent labeled Enzyme I showed a concentration dependent behavior. A Perrin plot of the emission anisotropy showed an inflection point at the temperature region of 6 to 16 $^\circ$C, the temperature region where the enzyme has been shown to exhibit the monomer dimer transition at this concentration range[34]. Moreover, emission anisotropy of the labeled enzyme varied as a function of pH, indicating that the monomer/dimer equilibrium is pH dependent. The higher anisotropy value was maintained between pH 7 and pH 7.5. Interestingly, Steeve and Waygood[45] reported that the activity of enzyme I was pH dependent with maximum activity at pH 7. These results indicate a good correlation between enzyme activity and the amount of dimeric form present. Nanosecond time-dependent emission anisotropy experiments also indicate that the monomer/dimer equilibrium is pH dependent.

Fig. 3 shows the effect of Mg^{2+} on the dimerization (Mg^{2+} is required for phosphorylation of Enzyme I from PEP). It appears that Mg^{2+} promotes dimer formation especially at the lower concentrations of Enzyme I. Fig.4 shows changes in tryptophan fluorescence as Ca^{2+} is added. The Ca^{2+} effect was diminished in the presence of Mg^{2+} or by addition of EGTA in the presence of Mg^{2+}, suggesting that the Ca^{2+} effect may be a replacement of Mg^{2+} by Ca^{2+}. This result suggests that tryptophan fluorescence is sensitive to subunit interaction and supporting data will be presented in the following section. Emission anisotropy of Enzyme I using tryptophan fluorescence indicates a decrease in the rotational correlation time, indicating subunit dissociation (Table 1).

D. Quenching studies of tryptophan fluorescence by energy transfer.

Enzyme I, which contains two tryptophan residues per monomer, exhibits two major decay times of 3.7ns and 7.3 ns and a small fraction of 0.6ns[46]. There is considerable spectral overlap between tryptophan emission and pyrene absorption, resulting in quenching of tryptophan fluorescence via energy transfer. When pyrene maleimide is attached specifically to Cys-575 of Enzyme I, the resulting enzyme contains two fluorescence energy donors (tryptophans) and a common acceptor (pyrene). Examination of the quenching pattern of tryptophan fluorescence can provide valuable information for assignment of the multi-decay components to individual tryptophan residues. Table 2 summarizes the

Table 1

Ca^{2+} EFFECT ON EMISSION ANISOTROPY OF ENZYME I.

$T(^{\circ}C)$	Ca^{2+}	Φ (ns)
23	None	117 ± 3
23	6.7mM	103 ± 2
23	20 mM	71 ± 2

Table 2

TIME-RESOLVED FLUORESCENCE DECAY PARAMETERS OF
NATIVE AND LABELED ENZYME I.

ENZYME I	τ_1	τ_2	τ_3	τ_4
NATIVE	.6	---	3.7	7.3
PM-LABELED	.66	1.85	3.8	7.1

#SH MODIFIED	$\%\alpha_1$	$\%\alpha_2$	$\%\alpha_3$	$\%\alpha_4$	$\%\alpha_{2+4}$
NATIVE	16	--	41	43	43
.5 SH	22	18	36	23	41
.8 SH	18	27	42	13	40
1.2 SH	17	41	40	2	43

decay parameters of Enzyme I as 0.5 to 1.2 moles of pyrene maleimide (per mole monomer) attached to Cys-575 at room temperature. An additional decay time (1.9ns) is observed in the pyrene-Enzyme I. It appears that the amplitude of the 7.1ns decay (τ_4) decreases as it is "exchanged", with the amplitude of the 1.9ns decay component (τ_2). On the other hand, the amplitude of 3.8ns does not seem to be diminished. This suggests that the 1.9ns component is a quenched decay of the 7.3ns species.

Neyroz et al.[34] reported that the two tryptophan residues of Enzyme I exhibit two distinctive decay-associated emission profiles. The 3.7 ns component is blue shifted compared to the 7.3 ns component. Moreover, the spectrum associated with the 3.7 ns component comprises 17% of total fluorescence intensity at 1°C (monomeric condition), whereas this component contributes 44% of total intensity under dimeric conditions (at 23°C). This suggests that one of the two tryptophan residues is relatively more sensitive to monomer/dimer association than the other tryptophan residue. In fact, the intensity of tryptophan fluorescence of Enzyme I dimer (at 23°C) is higher than the monomeric enzyme (at 1°C)[47]. This intensity change also accompanies a slight spectral shift, suggesting that conformational changes accompany subunit association. Moreover, the difference spectrum which reflects a blue-shifted emission spectrum is suggestive of the 3.7ns DAS. These results suggest that the two decay times originate from tryptophan residues in different unique environments. The only tryptophan close to a cysteine in the primary structure is Trp-498, which is close to cysteine-502[36,37]. Therefore, we would predict strong quenching between Trp-498 and an energy transfer acceptor bound to Cys-502, one of the "slow" reacting cysteines. In fact, the 3.7ns decay component was quenched to 0.4ns when the "slow" -SH groups were labeled with DTNB[46]. Considering these facts, Trp-498 appears to be in close proximity to the slow-reacting Cys-502 and is associated with the 3.7 ns decay component.

E. Kinetic measurments of monomer/dimer association
    As mentioned earlier, derivatization of four -SH groups of Enzyme I with DTNB resulted in inactive and monomeric enzyme, and consequently, tryptophan fluorescence is reduced by both resonance energy transfer (TNB is a good energy acceptor of Trp) and monomer/dimer associated conformational changes. Thus, it is of interest to closely examine the changes in tryptophan fluorescence to determine if these two processes can be

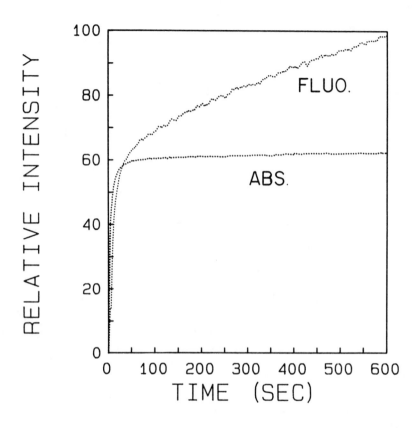

FIGURE 5:   The release of TNB from the DTNB modified
Enzyme I was measured by absorbance (OD_{412}) and Trp
fluorescence with excitation and emission wavelengths
of 295nm and 350nm respectively at pH 7.5 (room temperature). OD_{295} of the sample was 0.1.

distinguished during the TNB release from the DTNB modified Enzyme I by comparing the rate of TNB release (which can easily be measured by monitoring absorbance at 412nm) and changes in tryptophan fluorescence.

The results of this type of experiment are shown in Fig. 5. A 10 fold excess of dithiothreitol (DTT) are added at time zero. The release of TNB (as measured by absorbance) is essentially complete within one minute. In contrast, the kinetics of fluorescence recovery are biphasic and the reaction requires nearly two hours for completion. The rapid increase in the tryptophan fluorescence intensity is most likely associated with the loss of resonance energy transfer due to the fast release of TNB anion. The slow phase of fluorescence recovery is of interest because it may indicate conformational changes accompanying (or preceding) dimerization.

Supporting evidence for dimerization was obtained from kinetic steady-state emission anisotropy measurements with pyrene maleimide labeled Enzyme I[36,37]. Enzyme I was selectively and irreversibly labeled with pyrene maleimide at cys-575 (the C-terminal cysteine). The other three cysteine residues were reversibly labeled with DTNB. The time-course of dimerization was measured by monitoring changes in the emission anisotropy of the pyrene at $3^{\circ}C$ and $25^{\circ}C$ upon addition of dithiothreitol (simultaneous measurements of the vertical and horizontally polarized emission were performed with the aid of the T-configuration of a fluorometer)[36,47].

At $25^{\circ}C$, slow changes in anisotropy were observed (from 0.10 to 0.145). By contrast, only small anisotropy changes were seen at $3^{\circ}C$, a condition at which the native enzyme is predominantly monomeric. This confirmed the absence of large volume changes of Enzyme I upon TNB release. The rate of anisotropy changes seen in this experiment was very similar to the rate of tryptophan fluorescence recovery observed in the previous experiment (above).

Dimerization via a diffusion-controlled process is expected to occur on a second time-scale based on the concentration and size of monomeric Enzyme I. Thus, the slow rate actually represents a rate-limiting conformational change of the enzyme that allowing dimerization after transient encounters between monomers. Slow conformational changes are not unusual events in proteins. Our results favor the view that conformational changes precede the dimerization of Enzyme I.

III. FLUORESCENCE DECAY STUDIES OF HORSE LIVER ALCOHOL
DEHYDROGENASE (HLADH).

HLADH is a well characterized enzyme containing
two tryptophan residues (Trp 314 and Trp 15) and two Zn
atoms per subunit. One of the Zn atoms is at the active
site region and one is a structural Zn, involved in
maintaining the structure of the protein.

The heterogeneity of the HLADH fluorescence decay
process was characterized by Ross et al.[15], who used
both the wavelength dependence of decay times and the
selective quenching of Trp-15 by KI to assign decay
constants to each residue. Analysis of fluorescence
decay curves obtained as a function of emission wavelength made it possible to acquire emission spectra of
the individual Trp residues, an approach that had been
previously used by Brochon and coworkers in studies of
other enzymes[48].

Decay associated spectra (DAS) have also been obtained
by an alternative method. The mathematical formulation
of Knutson et al.[14] was largely motivated by the need to
collect data more quickly. The generalization of DAS to
a variety of overdetermination and association tools
was the primary goal of these recent studies. As an
example, QDAS[16,49] was a technique designed to test the
selective quenching of HLADH Trp-15 previously reported.[15,50,51] The excellent agreement seen between DAS
and independently gathered QDAS for this protein
confirmed both the hypothesis and the method.

A. Kinetic decay measurements

Data collection is a lengthy process in conventional single-photon counting nanosecond fluorometry.
This is one of the disadvantages often ascribed to this
technique. During a typical decay experiment one
usually collects over 10,000 counts at the peak of the
decay curve. It was demonstrated[16] that it is possible
to recover reasonably accurate decay times from the
decay curve with less than 300 counts at the peak.
Consequently, the problem of long collection times
attributed to the single photon counting technique is
more imagined than real.

In fact, the degree of correlation between decay
parameters is a more significant factor in the recovery
of lifetimes than the total number of counts collected.
The recent development of a global procedure for
simultaneous decay curve analysis[12,13] opened new possibilities for rapid decay measurements. With this data
analysis procedure reasonable decay times can be
recovered from a series of short decay curves collected
during biological and chemical processes[46,52]. This
kinetic decay measurement in which time is just another

axis for overdetermination is an extension of the "global approach". We will describe an example of "kinetic decay" measurements.

B. Kinetic decay studies of HLADH during acid denaturation.

Heitz _et al._[53] had previously used a multiple of probes to examine the acid denaturation of HLADH. The known assignment of the Trp decay pattern discussed above suggested that time-resolved studies could be used to better characterize the unfolding process. The first set of these so called "kinetic decay" measurements was performed by Walbridge _et al._[52], using flash-lamp technology. Thirty consecutive decay curves were collected at a rate of 1 minute per curve during the first half-hour of the acid denaturation of HLADH. Individual analyses of the data yielded (marginal) bioexponential decays. In the individual analyses, the only discernible trend was the loss of an amplitude. The entire data set was subsequently examined using global anlysis methods. The results suggested that the decrease in steady-state tryptophanyl fluorescence could be accounted for by the decrease in the amplitude associated with the longer of two decay components. However, the lifetimes recovered were 2 and 5 ns rather than the reported lifetimes of 4 and 7 ns for the native enzyme.

More recently, a synchronously pumped, mode locked, cavity-dumped dye laser capable of producing 20 psec (fwhm) pulses at a frequency of 4 MHz was used as a light source. The combination of the high repetition rate, high intensity light source with a computer controlled, high speed data collection device results in an instrument which has the potential for rapid collection times (a second or less). With this new technology, the earlier findings on denaturation were reexamined[54].

A pair of decay times (4 and 7 ns) matching the native decays and another pair (2 and 5 ns) corresponding to the denatured form were recovered from a global analysis. The complete domination of the "early decay curves" by native components that are lost as the denatured components (initially absent) appear is of interest. Walbridge _et al._[55] also extended kinetic decay to T-format, kinetic anisotropy decay ("KINDA") measurements using laser systems. Thus, rates of subunit dissociation and unfolding can be investigated with the aid of time-resolved fluorescence spectroscopy.

IV. SUMMARY

The fluorescence techniques described here show great potential in assigning decay components to specific tryptophan residues, examining protein conformational changes due to variations in solvent environments, separating ground- and excited-state heterogeneity of fluorophores, and determining factors essential for dimer formation and regulation of subunit interactions. The use of a series of fluorescence decay curve each obtained within few seconds, has been of value in better understanding the fluorescence of Enzyme I. We may anticipate that this technique might even more useful as it becomes possible to measure decay curves within a millisecond.

REFERENCES

1. Lakowicz, J.R., <u>Principles of Fluorescence Spectroscopy</u>, Plenum Press, New York (1983)
2. Cundall, R.B. and Dale, R.E., Eds, <u>Time-Resolved Fluorescence Spectroscopy in Biochemistry and Biology</u>, Plenum Press, New York (1983)
3. Badea, M.G. and Brand, L. in <u>Methods in Enzymology</u>, Vol. 61, Hirs, C.H.W. and Timasheff, S.N., Eds., Academic Press, New York (1979), PP 378-425.
4. Grinvald, A. and Steinberg, I.Z., <u>Analytical Biochemistry</u>, 1974, <u>59</u>, 583
5. Knight, A.E.W. and Selinger, B.K., <u>Spectrochimica Acta</u>, 1971, <u>27a</u>, 1223
6. Isenberg, I., Dyson, R.D. and Hanson, R., <u>Biophys. J.</u>, 1973, <u>13</u>, 1090
7. Helman, W.P., <u>Int. J. Radiat. Phys. Chem.</u>, 1971, <u>3</u>, 283
8. Gafni, A., Modlin, R.L. and Brand, L., <u>Biophys. J.</u>, 1975, <u>15</u>, 263
9. Amelot, M. and Hendrickx, M., <u>Biophys. J.</u>, 1983, <u>44</u>, 27
10. Amelot, M., Beechem, J.M. and Brand, L. <u>Biophys. Chem.</u>, 1986, <u>23</u>, 155
11. Valeur, B., and Moirez, J., <u>J. Chim. Phys. Chim. Biol.</u>, 1973, <u>70</u>, 500
12. Knutson, J.R., Beechem, J.M. and Brand, L., <u>Chem. Phys. Lett</u>, 1983, <u>102</u>, 501
13. Beechem, J.M., Ameloot, M. and Brand, L., <u>Analytical Instrumentation</u>, 1985, <u>14</u>, 379
14. Knutson, J.R., Walbridge, D.G. and Brand, L. <u>Biochemistry</u>, 1982, <u>21</u>, 4671-4679
15. Ross, J.B.A., Schmidt, C.J. and Brand, L. <u>Biochemistry</u>, 1981, <u>20</u>, 4361-4377
16. Brand, L., Knutson, J.R., Davenport, L., Beechem,

J.M., Dale, R.E., Walbridge, D.G. and Kowalczyk, A.A. in Bayley, P.M., and Dale, R.E., Eds., Spectroscopy and the Dynamics of Molecular Biological Systems, Academic Press, New York, (1985) pp 259-305

17. Knutson, J.R., Davenport, L. and Brand, L., Biochemistry, 1986, 25, 1805-1811
18. Davenport, L., Knutson, J.R. and Brand, L., Biochemistry, 1986, 25, 1811-1816
19. Davenport, L., Knutson, J.R. and Brand, L., Biochemistry, 1986, 25, 1186
20. Lofroth, J.E., J. Phys. Chem., 1986, 90, 1160
21. Beechem, J.M., Ameloot, M. and Brand, L., Chem. Phys. Lett, 1985, 120, 466
22. Meadow, N.D., Kukuruzinska, M.A., and Roseman, S. in Enzymes of Biological Membranes, Martonosi, A., Ed., 3, 523 (1984)
23. Branden, Carl-Ivar, Jornvall, H.E., Furugren, B., in The Enzymes, 11, 3rd Ed., Boyer, P.D., Ed., Academic Press, New York (1975) pp 103-190
24. Kundig, W., Gosh, S., & Roseman, S., Proc. Natl. Acad. Sci. U.S.A., 1964, 52, 1067-1074
26. Kundig, W., Kundig, D., Anderson, B. and Roseman, S. J. Biol. Chem., 1966, 241, 3243
26. Roseman, S. (1969) J. Gen. Physiol. 54, 138-180
27. Postma, P.W. and Roseman, S., Biochim. Biophys. Acta, 1976, 457, 213
28. Saier, M.H. and Roseman, S., J. Biol. Chem., 1976, 251, 6598
29. Saier, M.H. and Roseman, S., J. Biol. Chem., 1976, 251, 6606
30. Weigel, N., Waygood, E.B., Kukuruzinska, M.A., Nakazawa, A. and Roseman, S., J. Biol. Chem., 1982, 257, 14461-14469
31. Waygood, E.B., Biochemistry, 1986, 25, 4085-4090
32. Kundig, W., and Roseman, S., J. Biol. Chem., 1971, 246, 1393-1400
33. Weigel, N., Powers, D.A. and Roseman, S., J. Biol. Chem., 1982, 257, 14498
34. Neyroz, P., Brand, L., and Roseman, S., J. Biol. Chem., 1987, 262, 15900-15907
35. Beechem, J.M., Knutson, J.R. and Brand, L., Biochemical Soc. Trans., 1986, 14, 832
36. Han, M.K., Ph.D. Dissertation, The Johns Hopkins University (1988)
37. Han, M.K., Roseman, S., and Brand, L. J. Biol. Chem, 1989, submitted.
38. Knopp, J.A. and Weber, G., J. Biol. Chem., 1969, 244, 6309.
39. Weltman, J.K., Szaro, R.P., Frackelton, A.R., Jr.,

Dowben, R.M., Bunting, J.R., and Cathou, R.E., <u>J. Biol. Chem.</u>, 1973, <u>248</u>, 3173-3177

40. Holowka, D.A. and Hammes, G.G., <u>Biochemistry</u>, 1977, <u>16</u>, 5538-5545

41. Rao, A., Martin, L., Reithmeier, R.A.F., and Cantly, L.C., <u>Biochemistry</u>, <u>18</u>, 4505.

42. Gregory, J.D., <u>J. Am. Chem. Soc.</u>, 1955, <u>77</u>, 3922

43. Smyth, D.G., and Tuppy, H., <u>Biochim. Biophys. Acta.</u>, 1968, <u>168</u>, 173-180

44. Wu, C.W., Yarbrough, L.R., and Wu, Y.H., <u>Biochemistry</u>, 1976, <u>15</u>, 2863-2868

45. Waygood, E.B. and Steeves, T., <u>Can. J. Biochem.</u>, 1980, <u>58</u>, 40-48

46. Han, M.K., Walbridge, D.G., Knutson, J.R., Brand, L and Roseman, S., <u>Anal. Biochem.</u>, 1987, <u>161</u>, 479-486

47. Han, M.K., Knutson, J.R., Roseman, S., and Brand, L. <u>J. Biol. Chem.</u>, 1989, submitted

48. Brochon, J.C., Wahl, Ph., Charlier, M., Maurizot, J.C., and Helene, C., <u>Biochem. Biophys. Res. Comm.</u>, 1977, <u>79</u>, 1261-1271.

49. Knutson, J.R., Baker, S.H., Cappuccino, A.G., Walbridge, D.G. and Brand, L. <u>Photochem. Photobiol.</u>, 1983, <u>37</u>, S21

50. Abdallah, M.A., Biellmann, J.F., Wiget, P., Joppich-Kuhn, and Luisi, P.L. <u>Eur. J. Biochem.</u>, 1978, <u>89</u>, 397-405.

51. Eftink, M.R., and Jamerson, D.M. <u>Biochemistry</u>, 1982, <u>21</u>, 4443-4449

52. Walbridge, D.G., Knutson, J.R., and Brand, L. <u>Anal. Biochem.</u>, 1987, <u>161</u>, 467

53. Heitz, J.R., and Brand, L., <u>Biochemistry</u>, 1971, <u>10</u>, 2695.

54. Walbridge, D.G., Knutson, J.R., Han, M.K., and Brand, L. <u>Biophys. J.</u>, 1987, <u>51</u>, 284a

55. Walbridge, D.G., Han, M.K., Knutson, J.R., and Brand, L. <u>Biophys. J.</u>, 1988, <u>53</u>, 292a

8

Maximum Entropy Data Analysis of Dynamic Parameters from Pulsed-fluorescent Decays

By A. K. Livesey,[1,*] and J. Brochon[2]

[1]MRC LABORATORY OF MOLECULAR BIOLOGY, HILLS ROAD, CAMBRIDGE AND DEPARTMENT OF APPLIED MATHEMATICS AND THEORETICAL PHYSICS, SILVER STREET, CAMBRIDGE, U.K.
[2]LABORATOIRE POUR L'UTILISATION DU RAYONNEMENT ELECTROMAGNETIQUE, CENTRE NATIONAL DE LA RECHERCHE SCIENTIFIQUE, COMMISSARIAT À L'ENERGIE ATOMIQUE, MINISTERE DE L'EDUCATION NATIONALE, UNIVERSITE DE PARIS-SUD, F91405 ORSAY CEDEX, FRANCE.

1. INTRODUCTION

Often when trying to analyse dynamic parameters one is faced with having to invert the Laplace transform or sum of exponentials. This occurs in many fields such as Quasi-elastic light scattering, pulse-fluorescence decays, neutron spin-echo measurements and in kinetic studies of reactions. The Laplace transform is very ill-conditioned and although there have been many attempts to solve this problem, all of them give unwanted artifacts in the solution or have a limited validity or range of application. Maximum entropy methods, however, is set up to be immune from these artifacts and is completely general. In particular, it is equally capable of handling broad heterogeneous or sharp homogeneous species or mixtures of both.

In this paper we review the successful analysis of pulse-fluorescence data by maximum entropy[1] and give a quick presentation of current developments such as the analysis of rotational and flexural motions from polarised pulse fluorescence and the analysis of phase and modulation measurements of fluorescence. Maximum entropy has also been used to solve the Laplace transform in Quasi-elastic light scattering.[2-4]

2. PULSE-FLUORIMETRY

Pulse-fluorimetry is a technique used for measuring the heterogeneity of the chemical environment of a protein (or other

(*)Current address: Shell Research Limited,
Thornton Research Centre,
P.O. Box 1, Chester CH1 3SH.

biological or chemical entity) in solution together with its
rotational or flexural motions. This can help us to understand
both the structure of the protein in solution and its motion both
of which have a bearing on its function and the kinetics of its
interactions and reactions, both chemical and physical.

The measurement technique is in principle deceptively simple. An
infinitely sharp flash of polarised light is passed through a
solution of proteins which either has one or more natural
fluorescing centres (e.g. tryptophan) or has had one artificially
attached. The fluorescing centres are excited and after a
characteristic time depending on the local chemical environment
decays producing a photon of longer wavelength. In the meantime the
fluorescing centre may have rotated or flexed from its original
position so that the relative polarisation of the emitted signal is
now a function both of the angle between the absorption and emission
dipoles and the rates of rotation and flexion. Rather than
measuring the amplitude and phase of the emitted light we measure
the parallel $I_{//}$ and perpendicular $I_\perp$ components (from which the
former can be calculated).

In practice, the exciting lamp has a finite width compared to the
timescale of the phenomenon under study and so the sum of
exponentials has to be convolved with the (measured) excitation
profile. Thus:

$$I_{//}(t) = \frac{1}{3} E_\lambda(t) * [\int_0^\infty \int_0^\infty \int_{-0.2}^{0.4} \gamma(\tau,\theta,A)\ e^{-t/\tau}\ (1+2Ae^{-t/\theta})\ d\tau d\theta dA]$$

$$\ldots\ldots (1)$$

$$I_\perp(t) = \frac{1}{3} E_\lambda(t) * [\int_0^\infty \int_0^\infty \int_{-0.2}^{0.4} \gamma(\tau,\theta,A)\ e^{-t/\tau}\ (1-Ae^{-t/\theta})\ d\tau d\theta dA]$$

$$\ldots\ldots (2)$$

where E(t) is the temporal shape of the flash and $\gamma(\tau,\theta,A)$ are the
number of fluorophores with fluorescence decay τ, rotation time θ,
and initial amplitude of anisotropy A (related to angle between
absorption and emission dipoles). * Denotes a convolution with
time.

Typical lifetimes found in proteins range from a few hundred
pico-seconds to a few hundred nanoseconds. This requires
sophisticated measuring equipment, using either a pulsed-laser or
synchrotron as the light source.

Data analysis using the maximum entropy method

     Our problem is to determine the relative properties of species,
$\gamma(\tau,\theta,A)$ with decay constants (τ_i) rotational time constants (θ_j)
and initial anisotropics (A_k) having measured inevitably sampled,

noisy, and incomplete representations of the polarised emitted light
and flash profile. At the kernel of the data analysis we need the
inverse Laplace transform of the measured light deconvolved by the
flash $g(t)$. Inverting the Laplace transform is a very
ill-conditioned problem. As a result small errors in the
measurement of the fluorescence curve or the flash profile can lead
to very large errors in the reconstruction of $\gamma(\tau,\theta,A)$.

We can view this ill-conditioning, which leads to a multiplicity of
allowable solutions, in a different way. Consider the set A of all
possible shapes of the curves $\gamma(\tau,\theta,A)$ displayed as a rectangle in
Figure 1. We can calculate "mock" data sets from each $\gamma(\tau,\theta,A)$ in
turn and test whether they agree with the noisy data set.' All those
$\gamma(\tau,\theta,A)$ that agree with the data <u>within the experimental</u> <u>precision</u>
are bounded by a dot-dashed line. Some of these curves however are
unphysical (<u>e.g.</u> contain negative numbers), and can be rejected (by
the dashed line). The remaining subset of spectra (shown shaded) we
call the feasible set. Every member of this set agrees with the
data and is physically allowable.

Since the feasible set is infinite, we are forced to choose one
member and, we do so directly by maximising some function
$M[\gamma(\tau,\theta,A)]$ of the spectrum. This function is chosen so that it
introduces the fewest artifacts into our resulting distribution. It
has been proved by Livesey and Skilling[5] that only the
Shannon-Jaynes entropy function, S, will give the least correlated
solution in $\gamma(\tau,\theta,A)$. The function is defined as:

$$S = \sum_{l} p_l - m_l - p_l \log \frac{p_l}{m_l}$$

where $p_l = \gamma(\tau_i,\theta_j,A_k)$ and m_l is the model which encodes our prior
knowledge about the system before the experiment.

Practical data are noisy and it would be wrong to fit the data
exactly because the noise would then be interpreted as if it were
true signal. We chose to bound the feasible set by a chi-squared
statistic.

$$C = \sum_{k=1}^{M} \frac{(I_k^{calc} - I_k^{obs})^2}{M \sigma_k^2} \leq 1.0$$

where I^{calc}, I^{obs} are the kth calculated and observed intensities,
σ_k^2 is the variance of the kth point, and M is the total number of
observations. The summation is over both the parallel and
perpendicular components. Our programme automatically corrects
for the amount of scattered light present in the data and any
background.

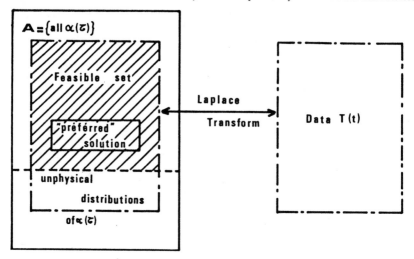

Figure 1. Diagram showing the set A of all spectra. Boundary of all spectra which agree with the measured data within experimental precision (-.-.-), boundary of the set of physically allowable spectra (---). The intersection of these two sets is the feasible set (hatched). Inside the feasible set, MEM will choose a "preferred" solution.

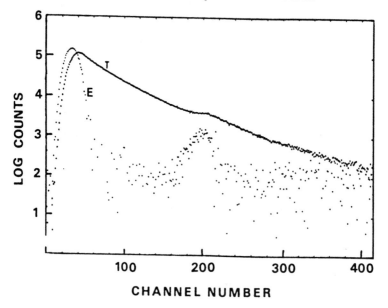

Figure 2a. Total fluorescence and excitation profiles for apocytochrome.

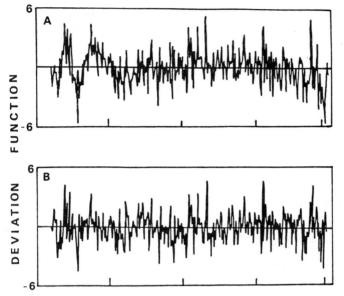

Figure 2b. Residuals showing excellence of fit between calculated and observed data when MEM is used to fit the data in figure 2a. Graph B follows on from A, ie the X-axis is at twice the scale of graph 2a.

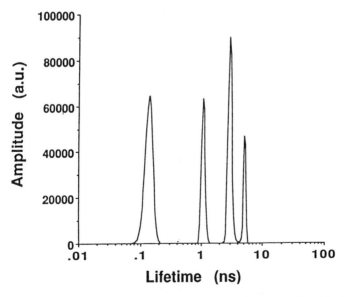

Figure 2c. Distribution of lifetimes in apocytochrome, as found by MEM from the data displayed in figure 2a.

The user may sometimes have some prior model[1,2] for the distribution but if not, it can be shown that the correct model is flat in log τ, log θ and cos λ space where:

$$A = \frac{3 \cos^2 (\lambda) - 1}{5}$$

The maximum entropy solution displays useful properties:

Because the log term, the distributions are all positive.

The solution only shows features (in particular the resolution of close peaks) if demanded by the data.

The spectra are smooth.

For data which are linear functions of the spectrum (this include Laplace transform) the Maxent solution is unique.

Provided the digitisation is sufficiently fine to show all relevant details, the shape of solution is independent of the number of points used to display it.

It is robust to noise.

<u>Total Fluorescence</u>

If we are only interested in the fluorescence decay constants τ we can considerably simplify the data by summing the parallel and (twice) the perpendicular components.

$$T(t) = I_{//}(t) + 2 I_{\perp}(t) = E * \int_{o}^{\infty} \alpha (\tau) \exp \frac{-t}{\tau} d\tau$$

where $\alpha(\tau)$ is the distribution of fluorescence decays given by:

$$\alpha(\tau) = \int_{o}^{\infty} \int_{-0.2}^{0.4} \gamma(\tau, \theta, A) \, dA \, d\theta$$

Note the variance of T(t) is:

$$\sigma_T^2 = \sigma_{//}^2 + 4 \sigma_{\perp}^2$$

T(t) can also be measured in one experiment by setting the polariser at the "magic" angle of 54.75°.

The entropy is also independent of θ and A and is given by:

$$S = \int \alpha (\tau) - m (\tau) - \alpha (\tau) \log \frac{\alpha(\tau)}{m(\tau)} \, d \, \tau$$

where the model m is flat in log τ space if the user has no knowledge of the τ values expected.

To illustrate the one-dimensional analysis we studied apocytochrome C(MW=12KD) which contains a single natural fluorophore tryptophan.

The measured total fluorescence and flash curves are displayed in Figure 2a and the reconstruction and residuals (showing the excellence of fit) are shown in Figure 2b and 2c. Four major components of the decay are clearly visible.

Further experiments as a function of temperature show a redistribution of lifetime components with the longest one almost disappearing at high temperature with the 1 ns component correspondingly increasing. This excited state lifetime heterogeneity may thus represent different environments for the TRP-59 fluorophore arising from different conformational sub-states of this protein in thermal equilibrium.

Polarised fluorescent decays

In order to test and illustrate maximum entropy's ability to analyse polarised fluorescent decays we show simultations. In each case we calculated theoretical parallel and perpendicular fluorescent data from "mixtures" of two fluorophores and added random Gaussian noise to the points with a variance equal to the number of photons detected. In each case we pretended to have a continued measuring until the maximum counts reached about 500,000. This large accumulation are quite feasible on the ACO synchrotron ring in LURE at Orsay, France.

In the first test the two fluorophores had a short decay (τ_1) together with a short rotational time constant. (θ_1) and a long decay (τ_2) with a long rotational time constant (θ_2). Both had an initial anisotropy amplitude of 0.3 (which is related to the angle between the absorption and emission dipole). In the second set the rotational time constants are swapped so that the two fluorophores have a short decay with long rotational time and long decay with short rotational time. Again both entities had the same initial anisotropy amplitude of 0.3. Finally to illustrate the full 3-dimensional analysis we returned to the decay and rotational constants of example 1 (*i.e.* short/short and long/long)

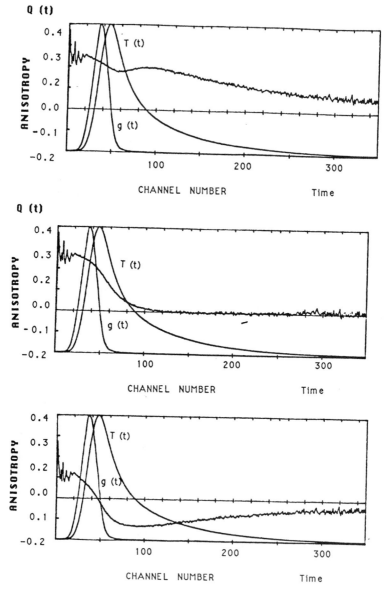

Figure 3. Computer generated data to test the MEM polarised fluorescence algorithm. T(t) is the total fluorescence, g(t) is the excitation profile and the unlabelled curve is the anisotropy R(t). In all cases there are two fluorescing species. In (a) the decay times and rotational time constants are (short/short, long/long, A1–A2–0.3), in (b) (short/long, long/short, A1–A2–0.3) and in (c) (short/short, long/long, A1–0.3, A2--0.15). Short decay – 1.1ns, short rotation – 0.6ns and long decay – 5.3ns, long rotation – 13.7ns.

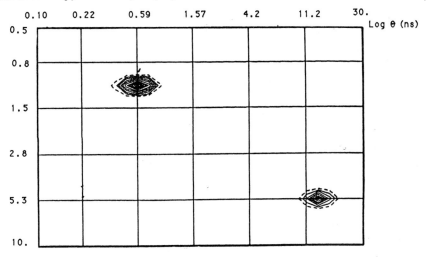

Figure 4. Maximum entropy analysis of computed data set shown in figure 3a (short/short, long/long, A1-A2-0.3).

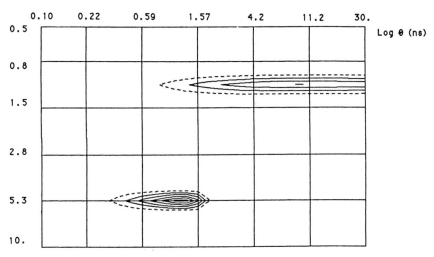

Figure 5. Maximum entropy analysis of computed data set shown in figure 3b (short/long, long,short, A1-A2-0.3).

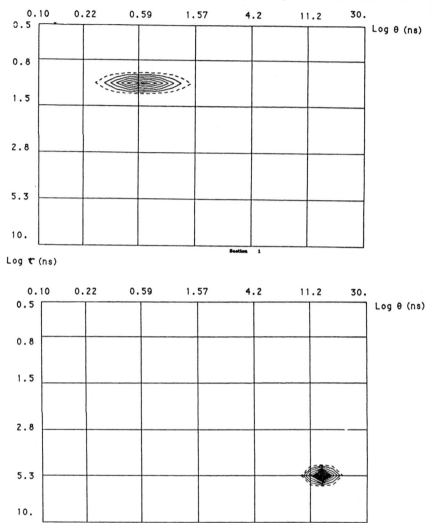

Figure 6. Maximum entropy analysis of computed data set shown in figure 3c
(short/short, long/long, A1=0.3, A2=-0.15). The first section is A=0.3 and
the second section is A=-0.15.

but in this case the initial anisotropy amplitudes are 0.3 and -0.1 respectively.

In Figure 3 we plot the flash, "measured" total fluorescence and the anisotropic decay R(t) given by:

$$R(t) = \frac{I_{//}(t) - I_{\perp}(t)}{I_{//}(t) + 2\ I_{\perp}(t)} = \frac{I_{//}(t) - I_{\perp}(t)}{T(t)}$$

If the flash is a delta function and the fluorescent species are in particular ratios then R(t) can analytically solved to give:

$$R(t) = \Sigma\ \beta_i\ e^{-t}/\theta_i$$

This separation (which only occurs in special cases) has led some workers to analyse R(t) as a sum of exponentials to determine the rotational time constants. Although this is not generally true, R(t) does give a pictorial representation of the quality of data available to determine the rotational constants.

In Figure 3a we can "see" that two true rotational decays are present. In Figure 3b only the shorter rotational decay is apparent. Although the longer rotational decay is present the total fluorescence quickly decays away so its presence is only weakly determinable. Finally in Figure 3c we see that there is a short rotational decay with a positive amplitude, and a long decay with a negative amplitude.

In Figures 4, 5 and 6 we display the maximum entropy analysis of these data sets. In Figure 4 we note the two entities are correctly determined as sharp peaks in their true positions. In Figure 5 the two entities have been recovered but we are less certain of their rotational time constant. In particular we are very uncertain of the rotational constant of the species with a long rotational constant because both the parallel and perpendicular measurements have decayed rapidly (due to the short decay τ_1) before we can get an accurate estimation of the long rotational decay.

In order to analyse the third simulated data set we used a full 3-dimensional analysis containing 3 sections with different initial anisotropics. We have only plotted the two extremal sections with amplitudes 0.4 and 0.2 respectively in Figure 6. The middle section was completely blank. As you can see maximum entropy has again correctly separated the two identities without further user intervention.

Polarised fluorescence of apocytochrome C protein
===

Data were measured at the synchrotron on a single tryptophane residue of apocytochrome in aqueous buffer at 20°C at the

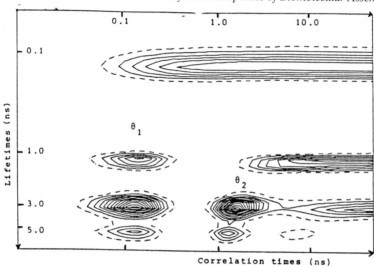

contour plot of a section through $\gamma(\tau, \theta, A)$ for A=0.258

PULSED FLUORESCENCE ANALYSIS OF A SINGLE TRYPTOPHAN PROTEIN

BY THE MAXIMUM ENTROPY METHOD: DISTRIBUTIONS OF LIFETIMES AND

ROTATIONAL CORRELATION TIMES OF APOCYTOCHROME C.

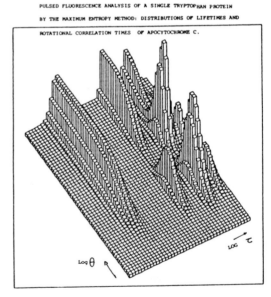

Figure 7. Result of the MEM solution of polarised fluorescence from Apocytochrome with A–0.258.

concentration of 2.5 mg/ml. The total scattering is shown in Figure 2 and the anisotropy profile is presented in Figure 7. The contour plot of a section through $\gamma(\tau,\theta,A)$ for A = 0.258 is presented in Figure 8. We can clearly distinguish four lifetime components centred at 0.15, 1.2, 3.14 and 5.4 ns as found previously in the one-dimensional analysis of T(t).

Along the θ axis we can see two high peaks from lifetime τ = 3.1 ns centred at θ_1 = 0.14 ns and θ_2 = 1.4 ns which reflect the major contribution to the depolarisation process. An identical minor contribution is given by the longest τ of 5.4 ns. However, the shortest lifetime (τ = 0.15) does not play any role in the fast motion (θ_1 = 0.14 ns). Its corresponding θ value remains uncertain due the weak contribution to the signal of a such short lifetime. Furthermore, the intermediate τ of 1.2 ns monitors the fast flexibility and the overall motion of the protein (MW=11 900) but is clearly not involved in the intermediate flexibility (θ_2).

In conclusion these examples fully demonstrate the ability of the maximum entropy method of analysis to resolve both the structural heterogeneity and complex protein dynamics from pulse fluorimetry data.

Acknowledgements

We are indebted to the technical staff at LURE for running the machines during beam-time sessions. P. TAUC and G. HERVE kindly supplied the labelled ATCase-AEDANS. We gratefully acknowledge M. VINCENT and J. GALLAY for measurements of polarised fluorescence decays of apocytochrome protein.

References

1. A.K. Livesey and J.C. Brochon, *J. Biophysical Society,* 1987, 52. 693.

2. A.K. Livesey, P. Licinio and M. Delaye, *J. Chem. Phys,* 1986 84, 5102.

3. A.K. Livesey, M. Delaye, P. Licinio and J.C. Brochon. *Faraday Discuss Chem. Soc.* 1987, 83, 247.

4. P. Licionio, M. Delaye, A.K. Livesey and L. Leger, *J. Physique (Paris)* 1987, 48, 1217.

5. A.K. Livesey and J. Skilling, *Acta Cryst A,* 1985, A41, 113.

Part II: Dynamics of Protein, Nucleic Acid, and Protein – Nucleic Acid Assemblies

9

An Approach to the Hydrodynamics of Elastically Linked Protein Oligomers

By A. J. Rowe

NATIONAL CENTRE FOR MACROMOLECULAR HYDRODYNAMICS (THE UNIVERSITIES OF NOTTINGHAM AND LEICESTER), UNIVERSITY OF LEICESTER, DEPARTMENT OF BIOCHEMISTRY, LEICESTER LEI 7RH, U.K.

INTRODUCTION

Macromolecular structures are far from being the static entities which the unwary reader might infer from study of illustrations in standard texts. Even such apparently compact structures as the smaller globular proteins are known from molecular dynamics calculations and by other methods to be in a state of constant vibration and sub-group oscillation or rotation on the picosecond time scale. Raman scattering, for example, yields information concerning local oscillatory groupings within a macromolecular solute, arising from bond stretching or rotation (1,2).

It is however customary to assume that the hydrodynamic properties of solutions (or dispersions) of such macromolecules require no consideration of intra-particle movement, since the methodology time-averages the parameters measured with respect to movement on the timescale in question. Relaxation of one part of the whole structure with respect to another, which will occur on a much longer time scale, is a different matter. Flexure in this case can be detected by specific labelling methods applied to the sub-structure, with consequent detection of the resultant depolarisation of fluorescence, fluorescence energy transfer, or whatever. Since these relaxations will be within orders of magnitude similar to the relaxation time of the whole particle in solution, it follows that they cannot be ignored in computations of the intensive properties of the solution.

Considerable interest has been focussed on the effect of flexure of particles on their solution properties (3). The use of temporally averaged particle conformations has provided a general approach, capable of describing both scattering and transport (e.g. diffusion, electrophoretic mobility) properties of flexible macromolecular solutes such as filamentous viruses (4,5).

A case which does not seem to have been studied however is one in which relatively large masses, whose intrinsic flexure can be considered to be negligible, are connected by elastic links. This may seem a problem of purely theoretical interest only. But in the case of the myosin filament from vertebrate skeletal muscle, there is strong reason for supposing that some half of the total filament mass is located in units attached to the filament shaft by linkages having elastic properties : further, the hydrodynamic properties of such filaments presents features difficult to interpret in terms of conventional explanations, including flexibility (6,7).
In an attempt to understand this and possibly other related systems, we have developed an approach towards a theory of such systems, which can be termed 'elastically linked oligomers' (ELOs). Elasticity differs from flexibility in that in the limiting case of an infinite flexibile linkage the connected units must still undergo motion (including density and concentration fluctuations leading to photon scattering) as spatially coherent entities; whilst with an infinitely elastic connection all coherence of motion is lost. This latter situation leads to a very simple treatment which can at its current state of development be used to give not a general solution to the problem but rather a quantitative account of the magnitude of the effects likely to result from the presence of elastic linkages.

AN APPROACH TO A THEORY OF ELASTICALLY LINKED OLIGOMERS

A harmonic oscillator model with temporal averaging of solute particle properties would seem to be an appropriate model, and it may well ultimately prove to be possible using carefully chosen approximations to consider at least simple systems in this manner. However a rigorous approach along these lines poses difficulties even at the level of 'order of magnitude' calculation (3). A very simple approach is possible however if one notes that the both the zero-elasticity and the infinite-elasticity cases are readily computable, and the assumption is then made that the (unknown and not computable) function relating the experimentally determinable parameter to the (finite) elasticity of the link is single valued in the latter argument.

One can then write, for the case of some measurable property which is a function of macromolecular parameters p,q, . . .

$$f(p,q,..)_e \;=\; \alpha f(p,q,..)_{e0} \;+\; (1-\alpha)\sum_{i=1}^{i=j} f_i(p,q,..)_{e\infty} \quad (1)$$

where the subscripts e, e0 and e∞ refer to the experimental, zero and infinite elasticity cases, the summation in the last term is over all the j elastically linked sub-units, and   is a dimensional-less weighting parameter (0 < α <1). Where multiple linkages are involved, it is assumed that they are equivalent, although an extension of equation (1) to multiple types of linkage would not be difficult.

The computation of the maximal effect resulting from elastic linkages by the use of equation (1) is trivial. Figure I shows the results of so doing for three simple models : two elastically connected spheres, and two elastically connected prolate ellipsoids (ratio of principal axes 5:1) for the cases of the zero-elasticity state being side-by-side or end-on.

<u>Figure 1</u>   The value of the translational friction, intrinsic viscosity and the amplitude scattering predicted for three models for the infinite elasticity case relative to the zero elasticity case in the linkage shown

Model	f, $D_T{}^{1}$	[η]	scattering
	1.59	1.00	0.5
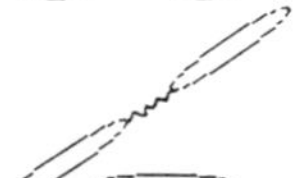	1.61	0.43	0.5
	2.33	1.85	0.5

The results are interesting. They show that the frictional coefficient increases and the scattering decreases, substantially in all three cases, but the intrinsic viscosity can either increase or decrease as a result of elasticity in the linkage. Clearly there is a potential effect of elasticity at a quantitative level readily detectable by conventional physical techniques. We have explored the application of this newly

developed theory initially to the case of myosin filaments :
future applications to other systems are however likely to be
fruitful.

MYOSIN FILAMENTS : A PROBLEM IN HYDRODYNAMICS

Myosin filaments from vertebrate skeletal muscle (6) are close
to 1.6 µm in length and have a shaft diameter of about 14 nm,
tapering at both ends. From the shaft protrude 294 pairs of 'S1
units' or 'heads', each of approximate dimensions 17 x 4 nm, and
packed into a 3 fold helical array (Figure 2).

<u>Figure 2</u>   Diagrammatic representation of the bipolar myosin
filament (top) composed of 294 myosin monomers (centre). Each
monomer contains 2 'head' or S1 units (enlarged scale, below)
attached via putative elastic linkages to the linear S2 units
and thence to the LMM units which pack into the filament shaft.

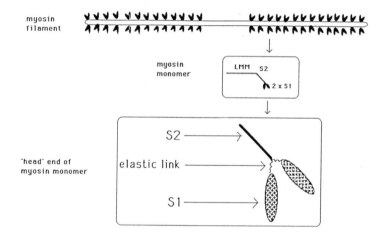

The total combined mass of the S1 units is similar to that of
the shaft. 'Synthetic' myosin filaments (SMFs) can be made by
polymerisation of the purifed monomeric myosin : such SMFs
are invariably less well ordered in their structure than native
filaments, and have a generally higher mass/unit length, and
often shorter overall length.
The sedimentation coefficient of either species (native or
synthetic) was noticed   (6) to be anomalously low as compared
with what might be computed from plausible models : <u>i.e.</u> the

frictional coefficient is unexpectedly *high*.   This can be illustrated in two ways.

(1) Mass/unit length via the treatment of Peacocke & Schachman

Peacocke & Schachman (8) showed that the sedimentation coefficient of linear particles is related to the semi axes of a prolate ellipsoidal model by

$$s_{20,w} = \{ 0.222(1 - \bar{v}\rho) b^2 \ln(b/a)\} / \bar{v}\eta \qquad (2)$$

where b,a are the short and long semi axes, v is the partial specific volume, and η is the solvent viscosity. From this relationship b is readily computed, being reasonably insensitive to an assumed value for b/a.

As pointed out by the original authors, this simple treatment is remarkably effective, even for particles where one might have expected a prolate ellipsoid to be a less suitable model than e.g. a cylinder (8,9). For the muscle filamentous polymer F actin, with a = 1000 nm and s = 40 S (9) and assuming b/a = 300, equation (2) yields

$$2b = 5.9 \text{ nm} : \text{ hence mass/36 nm}^* = 532 \text{ kD}$$
$$(\underline{cf} \; 13^* \; x \; 41^* \text{ kD} = 533 \text{ kD})$$

*F actin consists of 13 units of mass 41 kD in a 36 nm repeat of length.

Although the precise degree of agreement is fortuitous in this instance, the general point holds. A reason for the success of so simple an approximation is that any lateral swelling of a highly asymmetric particle results in cancelling effects on the frictional coefficient arising from greater particle volume balanced by lower asymmetry.

For native myosin filaments, use of equation (2) results in a prediction from the sedimentation coefficient (132 S) of 1.7 monomers per 14.3 nm filament length (cf the known value of 3 monomers per 14.3 nm).

(2) Computation of friction for a length equivalent ellipsoid

Emes & Rowe (6) used synthetic myosin filaments (SMFs) as a model system, and for a series of preparations of measured length and total mass computed the friction for a prolate ellipsoid of equal length and equivalent volume. For a range of preparation, the actual measured friction was higher by a factor of 1.7 than the predicted friction.

Thus myosin filaments whether 'native' or 'synthetic' show an anomaly in their hydrodynamic behaviour. This anomaly is very large in magnitude, can be modulated in its extent by divalent cation (11) and appears to have no parallel in any other system of extended particles studied to date. Explanations have been attempted by Emes & Rowe (6) and by Rowe & Maw (7). The former authors considered that the screw axis associated with the helical packing might cause extra friction : the latter authors that flow of solvent over the particle surface might induce cyclic movements of the S1 units. The former effect must be real: but can be shown to be rather small. The second effect may indeed occur : but it is far from obvious that even the algebraic sign of the effect on the friction can be predicted on this basis.

MYOSIN FILAMENTS CONSIDERED AS ELOs

From equation (1) the range of possible effects of an elastic linkage between the S1 units and the filament shaft has been computed :

....

TABLE 1

Measured parameter	friction	scattering
Range predicted	1 → 3.18	1 → 0.55
Experimental data	1.7	0.85
Implied value for	0.33	0.3 → 0.4

....

There is a prediction that the mass determined by scattering should be anomalously low. By contrast, mass determined by a combination of methods which eliminate the frictional coefficient (s & k_S; s & D from QLS) should be unaffected. The lowered value for the scattering and its modulation by divalent cation (11) has been demonstrated (Table 2) :

TABLE 2 Apparent relative* mass of SMFs by neutron
      scattering#

[cation]	relative* mass	R_G (nm)
0.1 mM Mg	0.85	14.0
3.0 mM Mg	0.88	14.1
2.0 mM Mn	0.95	13.8

*relative to independently determined values
#D L Worcester & A J Rowe, unpublished results

There is clearly in both cases (friction and scattering) a very good agreement between the magnitude of the anomalous effects observed and that predicted by equation (1).

CAN OTHER SYSTEMS BE CONSIDERED AS ELOs ?

Elastic linkages between oligomeric units are likely to be uncommon. However one can immediately predict that any polymeric structure into which myosin monomers are linked via their S1 units are likely to show anomalous hydrodynamic behaviour. We are currently studying one such system : native actin filaments with myosin monomers bound; which do indeed show a very large effect of the type described (C Brown, P Harrington & A J Rowe, current work). Further, from the relative insensitivity of the intrinsic viscosity to elasticity for globular proteins, it can be predicted that the shape function R ($= k_S/[\eta]$) could have a value *greater* than the theoretical maximum of 1.6 (12). Preliminary results indicate that this is almost certainly so for serum albumin; but there are probably rather other few cases where R > 1.6 (12,13).

CONCLUSIONS

There is very strong physiological evidence to indicate the presence of an elastic linkage (the 'series elastic component of muscle') between the S1 unit of myosin and the remainder of the myosin molecule linked into the filament shaft (14).   Hence our discovery that the presence such a linkage provides the basis for a theoretical approach to the anomalous hydrodynamics of these filaments offers a unified picture of both the physiology and the hydrodynamics of the S1   shaft

interaction.
It seems unlikely that the polypeptide structure of the S1 linkage in myosin, whatever it may involve, could be totally unique : and we are currently searching for other systems which on the basis of their hydrodynamics might be considered as ELOs. It may well also be possible to extend the present preliminary theory, to the level at which values for intermediate values of α can be computed for the simpler models.

ACKNOWLEDGEMENTS

I am grateful to the Medical Research Council and to the Science and Engineering Research Council for support of various aspects of this work.

REFERENCES

1. B. Chu, *'Laser Light Scattering '*, Academic Press, New York and London, 1974
2. R.C. Lord and N.T. Yu, *J. Molec. Biol.* , 1970, *50* , 509
3. S.P. Spragg, *'The Physical Behaviour of Macromolecules with Biological Functions '*, Wiley, Chichester U.K., 1980
4. E. Loh, E. Ralston and V.N. Schumaker, *Biopolymers* ,1979, *18* , 2549
5. E. Loh, *Biopolymers* , 1979, *18* , 2569
6. C.H. Emes and A.J. Rowe, *Biochim. Biophys. Acta* , 1978, *537*, 125
7. A. J. Rowe and M.C. Maw, *'Contractile Mechanism in Muscle '* (ed. G.H. Pollack and H. Sugi), Plenum, New York and London, pp 5 20.
8. A.R Peacocke and H.K. Schachman, *Biochim. Biophys. Acta* , 1954, *15* , 198
9. P. Johnson, D.H.Napper and A.J. Rowe, *Biochim. Biophys. Acta*, 1963, *74* , 365
10. A.J. Rowe, *'Techniques for determining molecular weight'*, in *'Techniques in Protein and Enzyme Biochemistry '* (ed. K.F. Tipton), Elsevier, Ireland, 1984
11. A. Persechini and A.J. Rowe, *Biopolymers* , 1984, 172, 23
12. A.J. Rowe, *Biopolymers*, 1977, *16* , 2595
13. J.M. Creeth and C.G. Knight, *Biochim. Biophys. Acta*, 1965, *102*, 549
14. J. Squire, *'The Structural Basis of Muscle Contraction '*, Plenum, New York, 1981

10

Laser Light Scattering Determination of Size and Particle Composition of δ-Endotoxin Crystals from *Bacillus thuringiensis*

By D. B. Sattelle, D. J. Langer, C. A. Haniff, and D. J. Ellar*

A.F.R.C. UNIT OF INSECT NEUROPHYSIOLOGY AND PHARMACOLOGY, DEPARTMENT OF ZOOLOGY, UNIVERSITY OF CAMBRIDGE, DOWNING STREET, CAMBRIDGE CB2 3EJ, U.K.
*DEPARTMENT OF BIOCHEMISTRY, UNIVERSITY OF CAMBRIDGE, DOWNING STREET, CAMBRIDGE CB2 3EJ, U.K.

1 INTRODUCTION

The polypeptide toxins of the Gram-positve bacterium Bacillus thuringiensis are highly toxic to insects[1] and are the basis of novel insecticides[2]. Strains of this micro-organism produce insoluble, phase-bright, cytoplasmic inclusions during sporulation[3,4]. These crystalline inclusions consist of one or more polypeptides (δ-endotoxins) which can be used as biological control agents against larvae of lepidopteran crop pests and dipteran (mosquitoes and blackflies) vectors of diseases. Different strains of Bacillus thuringiensis vary considerably in their toxicity towards various insect groups and this has been attributed variously to : (a) the presence of unique δ-endotoxins in the native crystal; (b) quantitative differences in the level of endotoxin production; (c) activation of the protein by the conditions (enzymes, pH) present in the gut of the host insect[5,6,7].

The variety Bacillus thuringiensis var. israelensis is active on Diptera, particularly mosquito and black fly larvae, but is much less effective against Lepidoptera[1]. Here we employ quasi-elastic laser light scattering (QELS) to investigate the physical properties of B.t. var. israelensis toxin. By comparing data with that obtained from B.t. var. kurstaki-HD1 toxin, which is active only against Lepidoptera, it should be possible to quantify any differences in the physical characteristics of these two toxin preparations. For the first time in studies on microbial toxins, we have examined particle-size distribution using the Provencher method of constrained regularization for analyzing scattered laser light signals[8,9].

Toxin size measurements for B.t. var. kurstaki-HD1 have previously been obtained using QELS[10]. In the present study we examine in detail hydrodynamic properties of purified toxins from two strains of Bacillus thuringiensis under alkaline conditions similar to those of the insect gut, and in the presence of SDS and dithiothreitol, with a view to assessing key factors regulating crystal association/dissociation.

2 MATERIALS AND METHODS

Toxins and chemicals

Frozen samples of purified toxins (0.1-0.2 mg ml^{-1} protein) from Bacillus thuringiensis var. kurstaki HD1 and Bacillus thuringiensis var. israelensis were obtained as described in detail elsewhere[6] and stored at -80°C. Samples were allowed to thaw, and were then sonicated for 15 min to remove aggregates. Unless otherwise indicated, aliquots of this material were diluted to a final concentration of 0.01 mg ml^{-1} protein in either distilled water (pH 7.0) or 50 mM sodium carbonate buffer. In both cases the medium used for suspending the crystals was filtered (Amicon UM10) to eliminate contaminants of M_r greater than 10,000. These diluted toxin samples were transferred to the quartz optical cuvette of a Coulter N4-MD submicron particle size analyzer. Following temperature equilibration (to 30°C) quasi-elastic light scattering (QELS) was performed. Dithiothreitol (DTT), sodium dodecyl sulphate (SDS) and sodium hydroxide were added directly to the optical cuvette containing the toxin sample from filtered stock solutions in order to achieve the required final concentration. Toxin crystals were also examined by phase-contrast light microscopy.

Quasi-elastic laser light scattering (QELS)

The mean diffusion coefficient, hydrodynamic size and polydispersity of biological macromolecules and supramolecular structures can be obtained from the measured autocorrelation function of scattered laser light[11,12]. For a sample of polydisperse scatterers, the correlation function is a distribution of exponentials, each characterized by its own decay rate

(Γ_i) associated with the diffusion coefficient (D_i) of each scattering species present. Using the moments (or cumulants) analysis[13,14], the autocorrelation function can be written as an expansion about $\overline{\Gamma}$, which is the mean of the distribution of decay rates weighted by the intensity scattered from each species. Second (μ_2) and third (μ_3) moments measure respectively the width and skewness of the weighted decay rate distribution. The intensity-weighted average diffusion coefficient ($\overline{D}$) can be obtained from the mean decay rate by

$$\overline{D} = \overline{\Gamma} / K^2$$

(where: K, the scattering vector, is given by $(4\pi n/\lambda_o)$ x $\sin^2 (\Theta/2)$; n is the refractive index of the medium; Θ is the scattering angle; λ_o is the wavelength of the incident light). From $\overline{D}$ the average hydrodynamic radius of an equivalent sphere can be calculated through the Stokes-Einstein equation. The normalized variance of the diffusion coefficient distribution ($\mu_2/\overline{\Gamma}^2$) provides a measure of the sample polydispersity.

A method of analyzing intensity autocorrelation functions developed by Provencher[15] utilizes a Fortran programme (CONTIN) developed at the European Molecular Biology Laboratory in Heidelberg, Germany[8]. This produces constrained regularization of linear equations enabling separation of particle size peaks which are not resolved by the moments (or cumulants) analysis. Its application to QELS data is well documented and enables the generation of particle size distribution histograms[9,16], which are able to reflect the multimodal particle size distributions of polydisperse samples. The programme makes no assumptions about the particle size distribution. Results are expressed as histograms of scattered light intensity against particle diameter.

3   RESULTS AND DISCUSSION

The hydrodynamic diameter of <u>Bacillus thuringiensis</u> var. <u>israelensis</u> toxin determined by QELS was approximately 750 nm in distilled water (pH 7.0, 30°C). The mean diameter ($\overline{d}$) increased slightly in 50 mM sodium carbonate buffer (pH 10.5, 30°C) while dropping dramatically following solubilization by dithiothreitol (DTT) (Table 1). No change in $\overline{d}$ was detected on addition of 1% sodium dodecyl sulphate (SDS), and this finding was supported by light microscopy, (Fig.1). When the measured diameters are compared to those obtained for <u>B.t. var. kurstaki</u> toxin, they are consistently slightly lower (Table 1), a result in good agreement with electron microscopical data[17].

<u>Table 1</u> Hydodynamic properties of δ-endotoxin crystals from <u>Bacillus thuringiensis</u>.   Mean   diffusion coefficient ($\overline{D}_{20,w}$), mean hydrodynamic diameter ($\overline{d}$), and polydispersity ($\mu_2/\overline{\Gamma}^2$) are recorded for the <u>var. israelensis</u> (Bt.i) and <u>var. kurstaki</u> (Bt.k) strains of <u>Bacillus thuringiensis</u> in different media. For $\overline{d}$, mean values are shown ± 1 S.E.M.

	$\overline{D}_{20,w}$		$\overline{d}$ (nm)		$\mu_2/\overline{\Gamma}^2$	
	Bt.i	Bt.k	Bt.i	Bt.k	Bt.i	Bt.k
Distilled water pH 7.0	5.67×10^{-9}	4.98×10^{-9}	758 ± 44	863 ± 38	0.23	0.20
Sodium carbonate buffer pH 10.5	4.88×10^{-9}	5.80×10^{-9}	881 ± 25	957 ± 32	0.21	0.18
Sodium carbonate buffer pH 10.5 +1% SDS	4.63×10^{-9}	4.05×10^{-9}	908 ± 51	1093 ± 48	0.29	0.34
Sodium carbonate buffer pH 10.5 +1% SDS and 25 mM DTT	1.06×10^{-8}	1.04×10^{-8}	404 ± 40	444 ± 85	0.23	0.62

Phase−contrast photomicrographs
of Bacillus thuringiensis δ−endotoxin

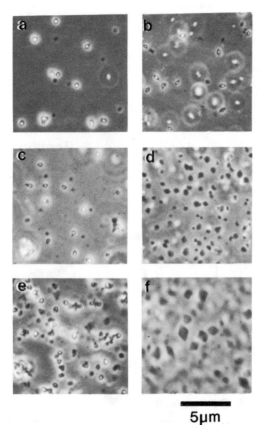

5μm

Figure 1   Phase-contrast photomicrographs of <u>Bacillus thuringiensis</u> δ-endotoxin crystals.   (a) <u>B.t. var. israelensis</u> and (b) <u>B.t. var. kurstaki</u> native toxin crystals in distilled water; (c) <u>B.t. var. israelensis</u> and (d) <u>B.t. var. kurstaki</u> toxin crystals in 50 mM sodium carbonate buffer (pH 10.5); (e) <u>B.t. var. israelensis</u> and (f) <u>B.t. var. kurstaki</u> toxin crystals in 50 mM sodium carbonate buffer (pH 10.5) containing 1% sodium dodecyl sulphate (SDS).   All micrographs obtained at 30°C.   Scale bar = 5.0μm.

Bacillus thuringiensis var. israelensis toxin was exposed to SDS and DTT and any changes in hydrodynamic properties were noted. Following the addition of DTT (25 mM) to toxin samples in buffer, a marked shrinking of the crystal and drop in the total scattered light intensity were observed indicating the onset of solubilization (Fig. 2). Complete solubilization was achieved approximately one hour after the addition of 25 mM DTT. By contrast, when SDS (1-10%) was added, no change in crystal size was detected, though some internal rearrangement of the crystal takes place as evidenced by a slight drop in light intensity. Also, following pretreatment with SDS (1%) more rapid DTT-induced solubilization was observed (Fig.3).

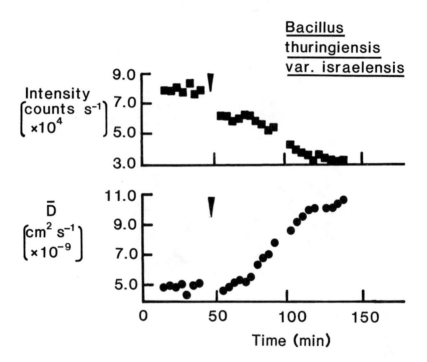

Figure 2 Changes with time of scattered light intensity and mean hydrodynamic diameter for Bacillus thuringiensis δ-endotoxin crystals. B.t. var. israelensis crystals were suspended in sodium carbonate buffer (pH 10.5) to which dithiothreitol (DTT) was added (to a final concentration of 25 mM) at the time indicated by the arrow.

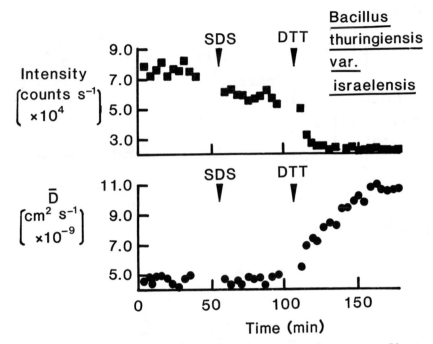

Figure 3   B.t. var. israelensis crystals were first suspended in sodium carbonate buffer (pH 10.5). Sodium dodecyl sulphate (SDS) at a final concentration of 1.0% and DTT (25 mM final concentration) were then added at the times indicated by the arrows.

The toxin of B.t. var. kurstaki showed a much more rapid response to DTT (25 mM) with complete solubilization within 12 min. Addition of SDS resulted in swelling of the toxin crystals (Fig. 1c, and Fig. 4). B.t. var. israelensis crystals did not swell following the addition of SDS but a significant drop in scattered intensity was detected. This indicates some loss or rearrangement of material making up the crystal without a change in hydrodynamic size. In this context it is of interest to note the results of recent electron microscopical observations[17]. These authors demonstrate the presence of a fibrous envelope surrounding the crystal of B.t. var. israelensis which delimits the crystal-like parasporal body. They also show that the envelope can survive solubilization of the proteinaceous contents of the parasporal body, leaving an empty shell.

This could account for the observed light intensity drop
without an accompanying change in hyrodynamic diameter.

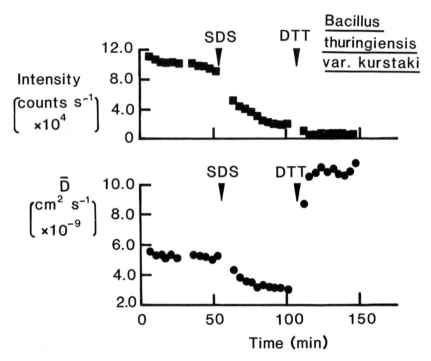

Figure 4 B.t. var. kurstaki crystals were suspended
initially in sodium carbonate buffer (pH 10.5), then
subjected to the same concentrations of SDS and DTT as
employed in experiments on B.t. var. israelensis
depicted in Fig. 3.   Here an SDS-induced swelling was
detected, that was not evident in the case of B.t. var.
israelensis.

DTT-induced solubilization of crystals from B.t.
var. israelensis indicates the importance of disulphide
bridges for maintaining their integrity.   Although B.t.
var. israelensis crystals showed no swelling following
SDS treatment, hydrogen bonding and hydrophobic
interactions probably play a role in crystal
organization since some internal rearrangement did occur
as evidenced by the heightened DTT response after SDS
treatment, as well as the light intensity decrease.

The degree of polydispersity (μ_2/Γ^2) was measured and a mean value found for each set of trials. For B.t. var. israelensis toxin, values ranged from 0.21±0.01 to 0.34±0.02 for all solution conditions tested (Table 1). Toxin crystals from B.t. var. kurstaki showed a much greater range of polydispersity values including a slight increase after SDS addition, and a marked increase in the present of DTT (Table 1). The kurstaki variety of Bacillus thuringiensis toxin was significantly altered by SDS and DTT and this is also reflected in polydispersity changes. Conversely, the B.t. var. israelensis toxin shows relatively little polydispersity change in different media, possibly attributable to the intact fibrous envelope.

Using the CONTIN programme[9] particle-size distribution was examined for the toxin crystals of B.t. var. israelensis. The vast majority (>90%) of samples examined were characterized by a bimodal distribution of particle sizes (Fig. 5). The predominant particle corresponded to the average size determined by unimodal analysis (800-1600 nm). Unimodal distributions were noted when the sample exhibited a lower than normal polydispersity value (<0.25). No significant change in particle-size distribution was observed after addition of SDS or DTT as the bimodal distribution continued to occur in greater than 90% of the trials.

Particle-size distributions were also obtained for B.t. var. kurstaki crystals and some differences from data reported for B.t. var. israelensis were observed. Though most distributions were bimodal (Fig. 5), unimodal distributions were observed in about 20% of the trials. Also, following addition of DTT (25 mM), more complex distributions were recorded in nearly 50% of the trials.

Laser light scattering has emerged as a sensitive probe of the hydrodynamic properties of Bacillus thuringiensis toxin crystals. Differences in particle size distribution and sensitivity to SDS treatment between different bacterial strains may reflect underlying differences in the chemical composition and intracrystalline hydrophobic/hydrophilic interactions. A detailed knowledge of crystal disassembly, and hence toxin activation, is essential for an understanding of the insecticidal mode of action of Bacillus thuringiensis δ-endotoxins.

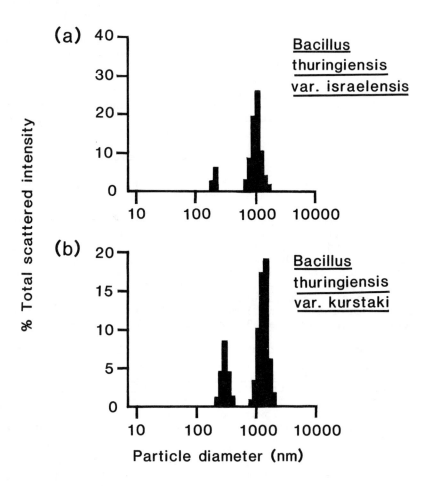

<u>Figure 5</u> Particle size distributions recorded from samples of <u>Bacillus thuringiensis</u> δ-endotoxin crystals. (a) <u>B.t. var. israelensis</u> crystals suspended in sodium carbonate buffer (pH 10.5); (b) <u>B.t. var. kurstaki</u> crystals suspended in sodium carbonate buffer (pH 10.5).

REFERENCES

1. H. Huber and P. Luthy, Bacillus thuringiensis
    δ-endotoxin: composition and activation. In: E.W.
    Davidson (ed) 'Pathogenesis of invertebrate
    microbial diseases', pp. 209-234. Allanheld, Osmun
    & Co., Totowa, New Jersey, 1981.

2. P.G. Fast, The crystal toxin of Bacillus
    thuringiensis. In: H.D. Burgess (ed) 'Microbial
    control of insect pest and plant diseases
    1970-1980', vol 2 pp. 223-248. Academic Press,
    London, 1981.

3. H.J. Somerville, Trends Biochem. Soc., 1978, 3,
    108.

4. H.T. Dulmage, Insecticidal activity of isolates of
    Bacillus thuringiensis and their potential for pest
    control. In: H.D. Burgess (ed.) 'Microbial control
    of insect pest and plant diseases 1970-1980', vol 2
    pp. 193-222. Academic Press, London, 1981.

5. T. Yamamoto and R.E. McLaughlin, Biochem. Biophys.
    Res. Commun., 1981, 103, 414.

6. W.E. Thomas and D.J. Ellar, FEBS Letts, 1983, 154,
    362.

7. M.Z. Haider, E.S. Ward and D.J. Ellar, Gene, 1987,
    52, 297.

8. S.W. Provencher, Computer Physics Commun., 1982a,
    27, 213.

9. S.W. Provencher, Computer Physics Commun., 1982b,
    27, 229.

10. D.B. Sattelle, C.A. Haniff, W.E. Thomas and D.J.
    Ellar, Biochim. Biophys. Acta, 1985, 840, 423.

11. B.J. Berne and R. Pecora, 'Dynamic light
    scattering', pp. 1-376. John Wiley & Sons, New
    York, 1976.

12. D.B. Sattelle, G.R. Palmer, M.C.A. Griffin and
    R.E.D. Holder, Med. Biol. Eng. Comput, 1982, 20,
    37.

13. D.E. Koppel, J. Chem. Phys., 1972, 57, 4814.

14. J.C. Brown and P.N. Pusey, J. Phys. D., 1974, 7, L31.

15. S.W. Provencher, Makromol. Chem., 1979, 180, 201.

16. S.W. Provencher, J. Hendrix, L. De Maeyer, J. Chem. Phys., 1978, 69(a), 4273.

17. J.E. Ibarra and B.A. Federici, J. Bacteriol., 1986, 165, 527.

11

Hydrodynamic and Fluorescence Analysis of a DNA Binding Protein and its Interaction with DNA

By G. G. Kneale

BIOPHYSICS LABORATORIES, PORTSMOUTH POLYTECHNIC, PORTSMOUTH PO1 2DT, U.K.

1 INTRODUCTION

DNA binding proteins are currently a major focus of attention in molecular biology; they include regulatory proteins such as bacterial repressors (eg *lac* repressor), and the transcription factors of eukaryotic cells (eg TFIIIA). The term in a more general sense includes proteins with a structural role such as the chromosomal proteins (histones) which fold DNA into nucleosomes. Other important DNA binding proteins are those that bind to single stranded DNA, which frequently play a role in replication . There is now evidence for a number of single stranded binding proteins that they can also bind to specific mRNA sequences and thus repress the translation of target proteins. In all these cases a prime objective is to determine the parameters that characterise the binding, and to obtain structural information concerning the protein-DNA complexes formed. In many cases the complexes are too large to be studied in detail by X-ray crystallography or NMR spectroscopy, especially as such proteins are often prone to aggregation. Even if this is not the case, complementary techniques are still required to characterise the binding.

In this article I will describe the the use of ultracentrifugation and fluorescence spectroscopy to investigate the properties of a single stranded DNA binding protein, the gene 5 protein of bacteriophage Pf1 [1], in order to illustrate the utility of these methods for the investigation of protein-DNA interactions.

The Gene 5 protein of filamentous bacteriophage

The gene 5 protein of filamentous bacteriophage fd plays a dual role in the replication of viral DNA. It binds to the single stranded viral genome, inhibiting the synthesis of the complementary strand [2]; it also represses the translation of the gene 2 protein, which is itself involved in replication[3]. The gene 5 protein of the related phage Pf1 is larger than its fd counterpart, being a 144 amino acid protein with a subunit molecular weight of 15,400 daltons. In the phage infected cell, it is found as a large complex (approximately $2x10^7$ molecular weight) consisting of a few thousand subunits of the protein bound to a single molecule of the phage DNA

[4] . The protein, surprisingly, is not released from the DNA by 2M NaCl but can be separated in the presence of 1M MgCl$_2$[1].

2 HYDRODYNAMIC STUDIES

Sedimentation Equilibrium.

Detailed hydrodynamic measurements have been done on the isolated protein to determine its molecular weight and shape. Sedimentation equilibrium experiments were carried out on a Beckman Model E analytical ultracentrifuge at a speed of 14,277 rpm using interference optics [5]. The data were analysed according to the method of Creeth and Harding[6]. The plot of ln c *versus* r^2 is almost linear (figure 1), indicating from the slope a molecular weight of 32,000 ($\pm$ 3000) daltons. However there is a small but significant deviation from linearity towards the bottom of the cell, indicating that there is some degree of aggregation. Nevertheless it appears that the major species in solution must correspond to the protein dimer. The weight averaged molecular weight over the whole cell was found to be 37,500 ($\pm$ 1000) daltons.

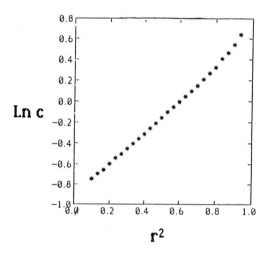

Figure 1. Sedimentation equilibrium at 14,277 rpm in the Beckman model E ultracentrifuge. The initial protein concentration was 0.4 mg/ml in low salt buffer.

Sedimentation Velocity.

The sedimentation profile of the protein solution was determined using an MSE centriscan with a rotor speed of 40,000 rpm at a variety of protein concentrations. Most of the protein sediments with a sedimentation coefficient s$_{20,w}$ of 2.6S (figure

2). However a significant proportion - around 30% - of the protein sediments very fast, with $s_{20,w}$ of around 35S. This component would not be seen in the sedimentation equilibrium experiment at the rotor speed used, as its molecular weight would be too great.

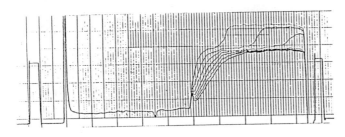

Figure 2. Sedimentation velocity scans of the protein at 0.67 mg/ml. Absorption was monitored at 278nm. The rotor speed was 40,000 rpm.

The 2.6S component we observe must correspond to the major species seen by sedimentation equilibrium. Taking the calculated molecular weight, M_r, of 30,800 and using the Svedberg equation:

$$S = M_r (1-\bar{v}\rho)/Nf$$

we can calculate the frictional coefficient f to be 5.19×10^{-8} , using a calculated value for $\bar{v}$ of 0.734 ml/g. The frictional coefficient for an anhydrous sphere of this size, f_o, would be 3.92×10^{-8} , thus giving a frictional ratio $f/f_o =1.32$ for the hydrated protein, fairly typical of globular proteins. Making an estimate of the hydration, one obtains a corrected translational frictional ratio of ca. 1.1 , which corresponds to an ellipsoid having an axial ratio of 2-3.

Salt induced aggregation.

There is biochemical evidence that aggregation of the Pf1 protein is enhanced in the presence of high concentrations of NaCl. This phenomenon has been investigated by sedimentation velocity runs on the protein in 2M NaCl. Under these conditions, there is no longer any hint of the 2.6S component and the major species has $s_{20,w} = 30S$. It is likely that this component is related to the 35S component observed in low salt buffers. Both show significant concentration dependence. Without knowing the shape we can not obtain an accurate molecular weight for the aggregated species. However, it must have a molecular weight of at least 1.5×10^6 daltons - more if it were significantly non-spherical. Electron microscopy [7] and X-ray fibre diffraction [8] indicate that the complex that the gene 5 protein makes with viral DNA is helical with a pitch of 45-55Å (depending on hydration) with 6 dimers per helical turn. It is possible that the aggregates may have a similar structure , corresponding to at least 50 dimers (8 helix turns).

3 FLUORESCENCE SPECTROSCOPY

The Pf1 gene 5 protein has a single tryptophan (Trp14) which dominates the fluorescence of the protein; no contribution is seen from the three tyrosines in the sequence. The excitation maximum occurs at 285 nm and the emission maximum at 343 nm. Fluorescence quenching experiments with Cs^+ and I^- indicate that the Trp14 is in a partially exposed negatively charged pocket [9]. On binding to DNA the emission peak shifts to 333 nm and the tryptophan becomes inaccessible to both ionic quenchers. These observations suggest that Trp14 is buried in a hydrophobic pocket when the protein is complexed to DNA. The evidence presented below suggests that it is the interaction between adjacent protein subunits on the DNA lattice, rather than any direct interaction with DNA , that is responsible for these changes to the tryptophan environment.

Fluorescence depolarisation

If a sample is irradiated with a short pulse of polarised light at its excitation maximum, the light becomes depolarised due to the tumbling of the molecules during the transition . The rate of depolarisation is then related to the rotational correlation time of the molecule. Rotational correlation times much larger than the lifetime of the excited state are difficult to measure as the light will not be significantly depolarised within the excited state lifetime. The anisotropy A(t) is defined as [10]

$$A(t) = \frac{[\,I_{\parallel}(t) - I_{\perp}(t)\,]}{[\,I_{\parallel}(t) + 2I_{\perp}(t)\,]}$$

where $I_{\parallel}$ and $I_{\perp}$ refer to the parallel and perpendicular components of the emitted light. If the molecule can be regarded as approximately spherical, then the anisotropy will decay exponentially with a rotational correlation time $\varnothing$.

$$A(t) = A(0)\exp(-t/\varnothing)$$

We have utilised fluorescence depolarisation to monitor DNA binding to the gene 5 protein . Figure 3 shows the anisotropy decay curve for the protein and for the purified gene 5 protein/viral DNA complex. The decay curves can be fitted by mono-exponential functions with $\varnothing$=19.2 ($\pm$ 0.5) ns and $\varnothing$=493 ($\pm$ 60) ns respectively. The latter value is subject to greater error for the reasons given above . The value of $\varnothing$ for the protein suggests that protein is predominantly dimeric, as observed by sedimentation. A hydrated sphere of this molecular weight would have a rotational correlation time of approximately 13 ns, and so the dimer must be somewhat anisometric in shape. The much larger value obtained for the complex with DNA corresponds to a rigid unit of around 15 dimers. The helical complex is known to have <u>ca</u>. 1800 protein subunits [8] but both the fluorescence data and the electron microscopy evidence point to substantial flexibility in the helix.

Oligonucleotide binding studies.

When the protein is bound to an excess of an octanucleotide, the rotational correlation time increases from 19.2 to 50.3 ns ; binding to hexadecanucleotides gives a complex with Ø= 85.0 ns (figures 3 and 4). Although it is impossible to deduce accurate shape or molecular weights from these figures, it is clear that the size of the complexes made increases with the length of the oligonucleotide. The binding curves resulting from titrations of the gene 5 protein with oligonucleotides (figure 4) allow the stoichiometry and binding constants of the system to be estimated [11]. The apparent stoichiometry of binding is close to 4 nucleotides per dimer, both with DNA and with oligonucleotides, and yet we know from biochemical data [1] and diffraction studies [8] that the two binding sites on the dimer each accomodate four nucleotides.

This apparent anomaly can be attributed to the cooperativity of binding, such that protein subunits bind initially along a single strand of DNA (or oligonucleotide) leaving the second site on the protein unoccupied (figure 5). As more DNA is added, the second sites will become filled but with no appreciable change in the size of the complex. Thus by this method, we would not see any significant change in rotational correlation time and thus we are effectively only titrating one of the two sites on the dimer.

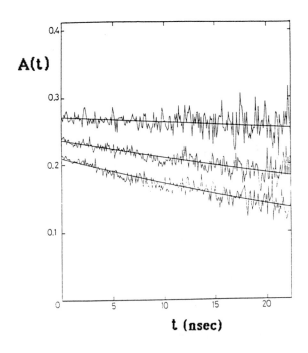

Figure 3. Anisotropy decay curves for complexes of the Pf1 gene 5 protein with viral DNA (upper), dT$_{16}$ (middle) and the octanucleotide d(GCGTTGCG) (lower) at saturation. The line of best fit for a monoexponential decay is drawn in each case.

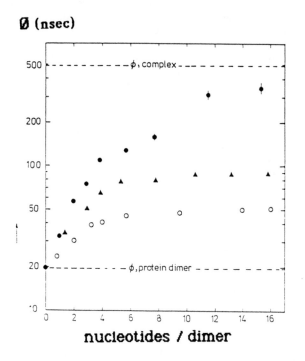

Figure 4. The rotational correlation time Ø of the gene 5 protein during titrations with viral DNA (•), dT_{16} (Δ) and d(GCGTTGCG) (o).

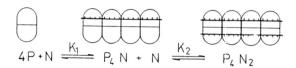

Figure 5. Schematic representation of the two binding modes of the Pf1 gene 5 protein. The single site mode will predominate when the concentration of DNA is limiting.

Spectral shifts and lifetimes.

We have also monitored the fluorescence lifetime and spectral shifts during titrations with DNA. Again, similar stoichiometries are observed. This would not be so if the tryptophan was experiencing a different environment due to a direct interaction with DNA. In fact when circular dichroism is used to monitor the titration, the "correct" stoichiometry of 8 nucleotides per dimer (ie 4 per subunit) is observed [12]. This suggests that the tryptophan fluorescence is affected by protein-protein contacts between adjacent dimers, and there is strong circumstantial evidence to support this [11]. For example, the spectral shift obtained when the protein binds to oligonucleotides of various lengths is found to be proportional to $(1-1/n)$, where n is the number of sites on the lattice, reflecting the number of intersubunit contacts made.

More detailed titrations performed in our laboratory with a variety of synthetic polynucleotides under a range of solution conditions in fact show that a variety of different binding modes are available. Additional information can also be obtained from performing reverse titrations, where increasing quantities of protein are added to a fixed amount of DNA [12]. In this case , as the DNA is in excess, both binding sites of the dimer contribute to the binding.

Fluorescence of salt-induce aggregates .

When the protein is measured in 2M salt, where aggregation is known to occur, the peak in the fluorescence spectrum shifts to lower wavelengths reminiscent of the DNA induced shift. This is in accordance with the proposal that the shift arises from protein-protein interactions. Measurements of the protein in high salt by fluorescence depolarisation show a rotational correlation time of 88 ($\pm$ 3) ns. An estimate of 4-6 subunits for the size of the aggregate is significantly smaller than the size of the salt induced aggregate estimated from its sedimentation coefficient. The rather different estimates for the size of the aggregates from fluorescence and centrifugation may be related to the physical basis of each method. In particular the presence of flexibility between subunits is likely to contribute differently in the two techniques.

ACKNOWLEDGEMENTS

I am grateful to my colleagues P.J.Morgan, S.E.Plyte, S.E.Harding M.L.Carpenter and R.Wijnaendts van Resandt for their contributions to much of the work described in this article.

REFERENCES

[1] G. G. Kneale, Biochim. Biophys. Acta , 1983, 739, 216
[2] J. S. Salstrom and D. Pratt, J. Mol. Biol. , 1971, 61 , 489
[3] W. D. Fulford and P. Model , J. Mol. Biol. , 1984, 173 , 211
[4] G. G. Kneale and D. A. Marvin, Virology , 1982 ,116, 53
[5] P. Morgan ,S. E. Plyte ,S. E.Harding and G. G. Kneale , Biochem. Soc.

Trans. (in press)

[6] J. M. Creeth and S. E. Harding , Biochem. J. , 1982, 205 , 635

[7] C. W. Gray, G. G. Kneale, K. R. Leonard, H. Siegrist and D. A. Marvin , Virology , 1982, 116, 40

[8] G.G.Kneale, R. Freeman and D. A. Marvin , J. Mol. Biol. , 1982, 156 ,279

[9] K. O. Greulich, R. W. Wijnaendts van Resandt and G. G. Kneale , Eur. Biophys. J. , 1985 , 11, 195

[10] P. Wahl, in "Biochemical Fluorescence Concepts" (eds. R.F.Chen and H. Edel hoch) ,1975, vol.1, 1

[11] G. G. Kneale and R. W. Wijnaendts van Resandt , Eur. J. Biochem., 1985, 149, 85

[12] M. L. Carpenter and G. G. Kneale , manuscript in preparation.

12

Mechanisms of Dynamic Ordering in DNA Solutions

By V. A. Bloomfield

DEPARTMENT OF BIOCHEMISTRY, UNIVERSITY OF MINNESOTA, 1479, GORTNER AVENUE,
ST. PAUL, MN 55108, U.S.A.

1 INTRODUCTION

DNA undergoes a variety of transitions in dilute or semidilute solution, from a monodispersed wormlike chain to a
compact and/or cooperatively ordered multimolecular complex. These transitions may be relevant in understanding
DNA packaging *in vivo*, and raise intriguing questions in
polyelectrolyte physical chemistry. They are often provoked
by a change in ionic strength or composition, but may also
involve addition of a neutral polymer or change in temperature. Those that we have studied include formation of a
very slowly diffusing extraordinary "phase" at millimolar
salt concentrations[1]; gelation of short (ca. 200 bp) DNA
molecules in solutions as dilute as 7-15 mg/mL [2]; condensation to toroidal and rodlike particles provoked by multi-
valent cations[3]; and aggregation at elevated temperatures
in the presence of moderate concentrations of divalent
metal ions[4]. This article will survey the phenomenology of
these transitions, and present our ideas about their mechanisms, including the possible coupling of secondary and
tertiary structure changes.

2 ORDINARY-EXTRAORDINARY TRANSITION

A striking example of apparent cooperative behavior in
dilute polyelectrolyte solutions is the observation by
quasielastic laser light scattering (QLS) of extraordinar-
ily slow diffusion coefficients in low salt solutions.
This phenomenon, which was first observed for 150 bp
mononucleosomal DNA fragments in our laboratory[1] has also
been found for dinucleosomal DNA fragments averaging 350
bp[5], polylysine, and polystyrene sulfonate. A review of

these systems is presented by Schurr and Schmitz[6].

The observations tend to follow a typical pattern. At high ionic strengths, the mutual diffusion coefficient D_m (measured by the decay rate of the QLS autocorrelation function) has the value D_o expected for a molecule of that size and shape. As the ionic strength is lowered, D_m rises and the total intensity of scattered light I falls in a reciprocal way predicted by theory. Below a critical salt concentration, the QLS autocorrelation function, which previously exhibited single-exponential behavior characteristic of a single diffusion process, begins to exhibit two-exponential behavior. The faster relaxation continues the upward trend of D_m, with steadily decreasing amplitude. At the same time, a much slower relaxation process with apparent $D \leq D_o/10$ appears. This slower process has an apparently constant D as the solution becomes more strongly interacting. This transition from normal to unusual scattering/diffusion behavior is termed the "ordinary-extraordinary transition". Although the ordinary-extraordinary transition is sometimes spoken of as a phase transition, there is no evidence of phase separation, for example the appearance of two layers of different densities. There is convincing evidence that individual molecules do not change their conformations. Our studies on DNA showed that the amplitude of the slow process remained essentially constant once it appeared; the decrease in total scattering intensity was due almost entirely to the fast component.

An important distinction in diffusion studies is between the mutual diffusion coefficient D_m measured by QLS and the self or tracer diffusion coefficient D_s measured by such techniques as fluorescence recovery after photobleaching or forced Rayleigh scattering. Experiments measuring D_s for both polylysine[7] and DNA[8] show that individual molecules move with diffusion coefficients not much different from D_o. This suggests that the slow process reflects some collective mode of motion.

The nature of the ordinary-extraordinary transition, and of the dynamical process(es) occurring in the slow phase, have been puzzles for a decade: no existing theory convincingly predicts or explains them. Existing theories of the coupling of polyion and small ion motions[9] predict a monotonic increase of D_m with decreasing salt. Stigter[10] suggested that the transition in polylysine may be explained by an extension of Onsager's[11] theory of the isotropic-anisotropic transition in solutions of rodlike molecules, with the effective rod diameter suitably enlarged to account for the increased range of interactions through the Debye-Hückel length. However, in our DNA solutions the transition occurs at a concentration about five-fold lower than predicted by Stigter's theory. Furthermore, the solution in which the slow relaxation is observed is not birefringent, as might be expected for an anisotropic phase. We have also explored the colloidal crystal theory[12,13], which seems appropriate for charged latex particles in deionized water, and found that the repulsive stabilizing forces do not seem sufficient for DNA solutions at the millimolar ionic strengths at which the transition is observed. An additional obstacle to a unified explanation of the ordinary-extraordinary transition in all polyelectrolyte solutions is the observation that 150 bp DNA fragments do not conform to the semi-empirical equation which appears to systematize data on polylysine and polystyrene sulfonate[14]:

$$\frac{C_p}{\sum_i C_i Z_i^2}\left(\frac{a}{Q}\right) = Z_s \tag{1}$$

where C_p is the equivalent concentration of charged polyion residues, C_i and Z_i are the molar concentration and valence of the i-th salt species, a is the average contour distance between charges along the chain, Q is the Bjerrum length, and Z_s is the valence of the counterion of added salt.

In view of these puzzles, we are currently in the process of accumulating additional experimental data over a wide range of presumably important variables, in order to obtain hints of directions in which theoretical explanations may be sought. The relevant variables can be expressed in terms of characteristic lengths of the system. These are the mean center-to-center distance between polymer molecules, which depends on concentration; the range of intermolecular forces, which depends on polymer size and on ionic strength through the Debye-Hückel length; and the inverse of the scattering vector q, which determines the length of the fluctuations and significant solution structures that may be observed at a given scattering angle and wavelength. The first two characteristic lengths are manipulated by changing DNA concentration and ionic strength. Intermolecular force range will also be sensitive to DNA fragment length; but this has been an awkward parameter to change, since large quantities of monodisperse short DNA - apart from mononucleosomal - have been hard to prepare. Some new purification techniques for plasmids should overcome this obstacle. Variation of the fluctuation wavelength q^{-1} is most difficult, since the char-

acteristic distance over which solution structures seem
likely to exist is about 500Å, too short for conventional
QLS measurements. However, this range is a suitable one
for small angle neutron scattering and small angle x-ray
scattering, and we intend to embark on such measurements
shortly, to attempt to detect Bragg peaks characteristic of
solution ordering.

   As noted above, the obvious theoretical approaches to
the ordinary-extraordinary transition have been unsuccess-
ful. There are, however, a couple of promising ideas yet
to be explored. Weissman[15] has demonstrated that small
amounts of polydispersity can lead to slow relaxations
associated with decay of the local polydispersity index.
Such polydispersity certainly exists in all systems investigated to date; e.g., in the mixture of 140 bp and 160 bp
fragments in mononucleosomal DNA preparations. The
theory[15,16] leads to

$$F(q,\tau) = (1-x)\ F^I\ (q,\tau)\ +\ x\ F^I{}_S\ (q,\tau) \qquad (2)$$

where

$$x \equiv 1 - \left\{ \overline{[b(q)]}^2 / \overline{b^2(q)} \right\} \qquad (3)$$

and F^I and $F^I{}_S$ are the "ideal" mutual and self dynamic
structure factors for monodisperse particles and $b_i(q)$ is
the scattering power of component i at wave vector q. The
analogy is made with neutron scattering, where the first
term in $F(q,\tau)$ is associated with coherent scattering, due
to fluctuations in the number density of particles regardless of type; and the second term is associated with inco-
herent scattering, due to fluctuations in the local polydispersity.

   This treatment assumes that particle sizes and interactions are identical, allowing decoupling of the two types
of fluctuation. For mononucleosomal DNA, this may not be a
bad approximation. Further development can be sought along
the lines suggested by Vrij[17], who used the Percus-Yevick
approximation to derive the static structure factor for
polydisperse hard spheres. Pusey and Tough[16] propose that
Vrij's approach could be extended to the dynamic structure
factor; and it would appear that rods could be treated as
well as spheres.

   A new approach which is even more attractive may come
from the work of Evans and coworkers[18,19] on diffusion
coefficients in ionic surfactant solutions. Experimentally,

the diffusion coefficient of SDS or tetradecyltrimethyl-
ammonium bromide in water is observed to pass through a
minimum in the vicinity of the CMC. This behavior seems to
be explicable only if the system is in the fast exchange
limit, in which the rate of association and dissociation of
micelles is faster than the diffusion rate $[(q^2 D_m)^{-1}$ in QLS
experiments].

This type of system is reminiscent of the "temporal
aggregate" picture proposed by Schmitz[20,21], which has not
yet been reduced to quantitative terms. To quote from
Schmitz et al[20]: "The slow mode in this interpretation is a
result of 'sharing' small ions between several polyions.
These 'temporal aggregates' are stabilized by a delicate
balance between the attractive forces arising from the
fluctuating dipole field generated from the sharing of
small ions by several polyions and the repulsive forces due
to the random Brownian motion and direct interactions
between polyions of like charge." This mechanism has some
physical plausibility, but its quantitative behavior has
not been worked out for a concretely specified model. We
have begun to do this, for a model in which the temporal
aggregate is likened to a surfactant micelle. The diffusion equations for this type of system have been carefully
worked out by Evans et al[18], and we are attempting to couple them to QLS behavior through the formalism developed by
Benbasat and Bloomfield[22] for nonionic systems. The systems of equations is extremely complicated and cannot be
solved analytically. However, numerical solutions should
indicate whether the temporal aggregate model gives the
proper sign and magnitude of the effect.

3 GELATION

In 1984, we[2] observed formation of gels in semidilute solutions of short, sonicated DNA fragments about 200 bp average length. These gels are quite different than the liquid
crystal phases studied at considerably higher concentrations by Rill and coworkers[23,24]. They are formed at concentrations well below the liquid crystal formation limit,
in the range 7-20 mg/ml. They are clear and optically
isotropic, and the scattering intensity of the system is
independent of the extent of gelation. The diffusion coefficient of the remaining mobile molecules, as measured by
QLS, is not affected by the presence of the gel. Gelation
is enhanced by Na^+ and Mg^{2+} and by lower temperature, and
is linear in the DNA concentration. Treatment with S1 nuclease did not affect gelation, indicating that it was not
due to association of dangling, single-stranded ends. Pel-

leting of the gel and electrophoretic analysis of the gel
and supernatant did not reveal any preferential incorporation of longer fragments in the gel.

These results are difficult to understand in light of
current theories of gelation[25], which would generally predict a higher-order DNA concentration dependence, and the
preferential incorporation of longer molecules. They also
raise intriguing questions about the structure of the gel
phase, which must be quite open and homogeneous, and about
the forces stabilizing the gel. Theoretical analysis[26] suggested that gelation might result from stacking of terminal
base pairs and entanglement of the linear multimers thus
formed. Perhaps in support of a base-stacking mechanism, we
have preliminary evidence that chaotropic anions such as
perchlorate and glutamate destabilize the gel relative to
structuring anions like sulfate or neutral ones like chloride. We hope to shed more light on these issues by studies, just beginning, that will more systematically vary DNA
molecular weight and use dynamic light scattering from
polystyrene latex particles as a probe of gel structure

4 CONDENSATION BY MULTIVALENT CATIONS

A typical unbroken viral or cellular DNA molecule, if
stretched out, would be millimeters or centimeters long.
If isolated in dilute solution, as a somewhat stiff random
coil, it would occupy a spherical volume of radius 1-10 mμ.
In fact, DNA in its functional biological environment -
packaged inside a phage capsid, a bacterial nucleoid, or in
chromatin - typically occupies only a small fraction of a
percent of the volume it would occupy in free solution.
This tight packaging presumably has many functions and consequences. These might include protection; selective expo-
sure and long-range spatial coordination of particular
regions for control of replication, transcription, and
recombination[27]; and coupling of secondary and tertiary
structure by torsion and bending.

The *in vivo* packaging of DNA, particularly in eukary-
otes, involves several levels[28]; different proteins appear
to mediate packing at each level. DNA packaging in phage
generally requires polyamines (putrescine and spermidine),
both basic and acidic polypeptides or small proteins, and
ATP as a source of energy[29]. Various mutants have been
found that allow at least some viable phage assembly in the
absence of one or another of these factors, but the impres-
sion is definitely created that a roughly equivalent number
of cationic charges are required to neutralize the DNA

phosphates. The acidic peptides may act by an excluded
volume mechanism[30], in the manner of the polymer and salt-
induced (psi or ψ) transition[31-34]. The amount of DNA
packaged is determined by the size of the preformed capsid.
The manner of winding DNA within phage heads is still not
clear after several studies[35-37], but toroidal arrays, or
figure-eights like twisted skeins of yarn, are often seen
after lysis of capsids[38], and some sort of reproducible
packing geometry seems to hold.

An understanding of the folding and tight packaging of
bare DNA, interacting only with itself, water, and small
ions, is fundamental to understanding of these higher levels. Several types of model systems have been developed
that enable study of condensation of DNA provoked by relatively simple environmental variables. These include condensation in the presence of polyamines or other multi-
valent cations[3,39-42], aggregation by divalent cations at
elevated temperatures[4] or in alcohol-water mixtures[3], and
psi condensation[31-34].

Condensation of DNA provoked by trivalent cations typically results, if the DNA concentration is very low, in
discrete toroidal particles that have roughly 400-500 Å
outer radius and 150 Å inner radius. Most frequently such
experiments have been performed with high molecular weight
DNA, in which only one molecule is sufficient to make up a
toroid. We have done similar experiments[43], using EM and
QLS, with short DNAs: 2700 bp linearized pUC12 plasmids and
1300 bp half-molecules (more precisely, an equimolar mixture of 1100 bp and 1500 bp fragments) produced by a single
restriction cleavage. Remarkably, the toroids formed by
these short fragments have almost exactly the same size,
which is also the size of toroids formed by much longer
molecules. The shorter the DNA, the more fragments are
incorporated in the toroid (about 13 linearized plasmids,
26 half-molecules). This shows that size determination of
toroids is an intrinsic property of DNA, independent of
length. Uncut, covalently closed circular plasmids, on the
other hand, typically form somewhat smaller toroids, with
central holes very small or absent.

Electron microscopy also shows that higher aggregates,
produced at higher DNA concentrations or after extended
periods, are not random aggregates of double helical
strands, but more often tend to be discrete toroids linked
together by a few bridging strands[43]. This picture is
reinforced by a pattern recognition-cluster analysis[44] of
the QLS and total intensity scattering of condensed pUC12
DNA, which shows three distinct states: random coil,

collapsed coil, and aggregrated chains. These results
suggest that DNA condensation and aggregation occur through
highly structured intermediates, rather than the random
aggregates implicit in current theories[45,46].

In fact, if DNA condensation is viewed as the association of a moderate number of molecules, leading to a well-
defined most probable aggregation number and dispersion
about that number, a similarity to micelle formation
becomes apparent. The theory developed by Tanford[47] enables
calculation of the mole fractions of particles as a function of aggregation number, if the dependence of particle
free energy on aggregation number is known or postulated.
In order for a maximum in the mole fraction to be observed,
repulsive and attractive free energy components must have
different dependences on aggregation number.

As shown by Tanford[47], the mole fraction of monomer
incorporated in a particle containing n monomers is

$$\ln X_n = \ln X_1 + \ln n - n\Delta G°(n)/k_B T \qquad (4)$$

where X_1 is the mole fraction of free monomer. $n\Delta G°(n)$ is
the free energy change on incorporating n monomers into the
aggregate. Unfortunately, this free energy function is not
known, making impossible the convincing calculation of size
distributions at this time.

The sorts of terms that should be included in a proper
theory are reasonably clear. They include contributions
from DNA bending and/or kinking, and entropy loss upon condensation of the random coiled DNA molecule into a compact
structure; these terms will be linearly proportional to the
amount of DNA incorporated in the aggregate, that is to the
product of n and the number of base pairs L[48]. Terms representing electrostatic repulsions arising upon near approach
of negatively charged helical segments[49], and water structure interactions of the sort discussed by Rau et al.[50], are
expected to vary somewhat more strongly than linearly with
the amount of DNA incorporated into the condensed particle.
The attractive forces might be expected to be proportional
to the square of the amount of DNA for small extents of
aggregation (since attractive interactions are likely to be
short-ranged and pairwise additive), and proportional to
the first power of nL for larger particles (since in this
limit the attractive interactions will be proportional to
the amount of new DNA incorporated).

Our explorations of possible dependences of $\Delta G°(n)$ on
n and L have found some functions that give distributions

in reasonable accord with experiment, but they have as yet
no basis in fundamental theory. Whatever form is assumed,
it is clear that the free energy of condensation per base
pair is only 10^{-2} to 10^{-3} of the thermal energy k_BT,
reflecting the delicately balanced and highly cooperative
nature of the condensation process.

While toroids are most often observed upon condensation of higher molecular weight DNA, we observe significant
fractions (often 50% or greater) of short rods in the lower
mol wt DNA preparations. Different parts of EM grids have
different ratios. Remarkably, the length and thickness of
the rods is almost exactly the same as the mid-line circum-
ference and thickness of the toroids observed with the same
molecules. This suggests that size determination is the
basic factor in DNA aggregation, with shape of the aggregate being a second-order effect perhaps depending on local
solution conditions. Since the length of a rod is less
than the extended length of a 1300 bp DNA molecule, the
molecules must be folded back 2-3 times, perhaps accounting
for single-strand nuclease-sensitive regions observed by
others. Very few rods are observed when covalently closed
circular plasmid DNA is condensed.

5 AGGREGATION BY DIVALENT CATIONS

In our work on condensation of DNA by polyamines[3], we used
an extension of Manning's counterion condensation theory[51]
to show that DNA condenses when about 89% of its charges
are neutralized by counterion binding. We observed that
for divalent cations, the necessary degree of neutraliza-
tion only occurred in the presence of alcohol. We
attributed this to strengthening of electrostatic forces
through a lowering of the dielectric constant ε. However,
first Rau and Parsegian [personal communication] and then
we[4] showed that DNA aggregated in the presence of 30-50 mM
concentrations of divalent metal ions at somewhat elevated
temperatures. Since the product $T\varepsilon(T)$, which enters the
counterion condensation theory, changes very little with
temperature T, we were led to conclude that divalent metal
ion-induced aggregation of DNA, in the presence of alcohols
or at elevated temperature, is the result of some more
specific chemical and/or structural change.

A search of the literature[4] reveals several reports
over the years that divalent metals can induce DNA condensation (aggregation) at elevated temperatures. Because of
the attendant complications of precipitation and turbid-
ity, the conditions giving rise to aggregation were usu-

ally noted by the workers and thereafter avoided. However, we feel that this phenomenon is valuable to shed light on general mechanisms for DNA condensation by short-range interactions.

Three models can be proposed to explain why DNA aggregates at elevated temperatures in the presence of moderate concentrations of divalent metal salts[4]: (1) complete melting followed by aggregation of the separated strands, (2) partial melting followed by both inter- and intra- molecular aggregation of the denatured regions, and (3) some type of aggregation involving only intact duplexes. Some consequences of these models along with supporting or contradicting evidence are discussed below.

Model 1: Complete melting

Over the years, many have reported the lowering of DNA melting temperatures in the presence of divalent cations. The order seen by Eichhorn and Shin[52] for helix destabilization by divalent metals is the same as that seen by us for precipitation effectiveness[4]. Thus, it may be that these metals exert their effect by a drastic reduction of helix stability, causing full-scale dissociation of the helix. If this is the case, it is not at all surprising that aggregation would result, since single stranded DNA and RNA aggregate in the presence of divalent metals, even at 4°C. Our S1 nuclease digestions and reversibility results argue against such a model, since completely melted calf thymus DNA cannot be readily renatured. Also, this model does not explain why the alkaline earths are effective as precipitants, since they are not believed to interact strongly with the bases or to drastically reduce T_m.

Model 2: Partial melting

If melting is incomplete, then in some regions of the DNA the two strands will remain in register and thus permit rapid renaturation upon the removal of metal. This behavior was observed in our experiments upon addition of EDTA and was also proposed for DNA melted in the presence of Zn^{2+} and Cu^{2+} [53-55]. Yurgaitis and Lazurkin[56] have shown that DNA gradually aggregates if the temperature is held just below the beginning of the melting transition. They attribute this effect to the presence of transiently melted regions, which aggregate in the presence of Mn^{2+} ions. Such progressive aggregation as the temperature is increased is also seen for other divalent metals[57]. As in Model 1 the order of effectiveness may be expected to follow that seen by Eichhorn and Shin[52].

A parallel can be drawn between our DNA aggregates and the P-form of DNA, obtained in solutions of high alcohol content by Johnson and coworkers[58-60]. Even though the conditions of formation are quite different, some characteristics of P-form DNA are similar to our system. Both types of DNA are aggregated and both can be induced to rapidly disaggregate. In the case of P-DNA, the disaggregation and renaturation occur despite the almost complete denaturation of P-form DNA (insofar as denaturation is indicated by lack of base pairing and stacking). Thus, rapid reversibility cannot be taken as proof of the absence of denaturation, nor does lack of base pairing and stacking prevent formation of ordered aggregates.

Model 3: No melting

A third possibility is that the DNA self-associates while the helix remains intact. This is apparently the case for cation- and Ψ-condensed DNA, as evidenced by CD[40], X-ray diffraction[34], and melting studies[61]. The low aggregation temperature that we observe for some metals, and the absence of hyperchromicity, are consistent with this model. Left unexplained by this model is the attractive force(s) responsible for aggregation. Purely electrostatic mechanisms, such as correlated fluctuations in the ion atmospheres surrounding nearby helices, cannot explain the cation and anion specificity of aggregation.

While type II metals may cause melting of DNA (or at least cause UV spectral changes characteristic of the hyperchromic effect) at metal/phosphate ratios near unity, our experiments and others suggest that the DNA may not be melted at higher metal/phosphate ratios. To explain this, we[4] have developed a model of a transition from B-form helix to an alternative helix or coil form X with bases accessible for site binding by z-valent metal ions M^{z+}. It is supposed that X has spectral properties similar to thermally denatured DNA, but has a charge spacing b_X which may be different from the 4.1Å normally assumed for single-stranded DNA. The calculation uses an adaptation of the theory by Record et al.[62] of the ionic strength dependence of the thermal helix-coil transition, where divalent counterions are in excess and determine both the counterion condensation and the ionic strength.

The equation describing the dependence of melting temperature T_m on salt concentration is[62]

$$dT_m/d\ln a_\pm = \alpha\beta'\Delta\psi \qquad (5)$$

where α is a nonideality correction factor near unity, and
ß' is proportional to the enthalpy of melting. Combining
specific binding, counterion condensation and Debye-Hückel
screening effects, the thermodynamic differential ion
association parameter $\Delta\psi$ (the effective number of ions
liberated in the B-X transition), is

$$\Delta\psi = [b_X^\circ/(1-zr) - b_B]/2zb_j - r \qquad (6)$$

where b_X° is the linear charge spacing of the X form in the
absence of site-binding, b_B is the charge spacing in B-DNA,
b_j is the Bjerrum length, z is the metal ion valence, and r
is the number of ions bound per phosphate.

In agreement with experiment, numerical evaluation of
Equation (6) predicts an intermediate range of salt concentrations at which the X form will be more stable, if its
charge density is not much less than that of the B form
(i.e. $b_X^\circ \approx 2\text{Å}$). This is consistent with the Eichhorn and
Shin[52] results, if the optical change results from un-
stacked bases in the X form, but the strands are still
intertwined so rapid renaturation is possible upon removal
of metal ion. However, at the high salt concentrations at
which our experiments were conducted, the double-stranded
form should again be more stable.

Of the three models presented for the aggregation of
DNA by divalent metal ions at elevated temperatures, we
have ruled out the first one, which entails complete strand
separation. At this point, however, we cannot choose
between Models 2 and 3. The Ψ- and polyamine or hexamine
cobalt(III)-induced condensations are precedents for Model
3. In view of the observed metal specificities and the
previous work of others, we must consider model 2 a distinct possibility. A possible mechanism that includes
aspects of both models 2 and 3 is the formation of some
modified but regular secondary structure of DNA which has a
strong tendency to aggregate.

It is therefore intriguing to speculate that divalent
metal ion-induced aggregation of DNA may be related to formation of some form of DNA, such as Z-DNA or P-DNA, which
largely maintains integrity of the double-helical backbone
winding, with a linear charge density not much different
from that of B-DNA, but which has unpaired and unstacked
bases exposed to the solvent environment. For example, Z-
DNA has a spacing per phosphate of 1.9 Å [63]. Such a form of
DNA could show spectral changes characteristic of denaturation, bind to metals, and engage in interhelix base pair-

ing. This would account for many of the observations in this and other studies on metal-induced aggregation of DNA. It would also connect with observations on cruciform formation: Sullivan & Lilley[64] have found that the order of effectiveness of various divalent metal ions in causing cruciform extrusion is similar to that we have observed for aggregation.

6 AGGREGATION AND CHANGES IN HELIX STRUCTURE

These results on DNA aggregation by divalent metal ions, on the properties of P-DNA, and on condensation of alternating purine-pyrimidine copolymers by trivalent cations described below, suggest the intriguing possibility that DNA aggregation may generally be due to the formation of non-B helical forms of DNA even without overt strand separation.

Many of the ionic conditions that lead to the B-Z transition in alternating pyrimine-purine copolymers (very low salt with multi-valent cations in water, or divalent cations at higher temperatures or in water-alcohol mixtures) are also those that cause condensation or aggregation[65-68]. It has been observed by several groups that Z- or other non-B helical forms are often produced coincidently with collapse in such alternating copolymers[69-71], and that the Z* form of poly(dG-dC), produced with transition metal ions in EtOH or at high T, has strong tendencies to aggregate[72,73].

In our own work, we have used a combination of spectroscopic techniques, QLS, and electron microscopy to show that the B to Z transition in poly(dG-m^{5}dC)·poly(dG-m^{5}dC) is accompanied by extensive condensation of the DNA in both low and high ionic strength buffers[74]. At low concentrations of NaCl (2 mM Na$^+$), an intermediate rodlike form, which exhibits a CD spectrum characteristic of an equimolar mixture of B and Z forms, is observed. This is produced by the orderly self-association of about four molecules of the polymer after prolonged incubation of a concentrated solution at 4°C. On addition of 5 μM Co(NH$_3$)$_6$$^{3+}$ the CD spectrum of the intermediate changes to that of the Z form, which is visualized as a dense population of discrete toroids on an EM grid stained with uranyl acetate. On the other hand, addition of NaCl to a solution of poly(dG-m^{5}dC)·poly(dG-m^{5}dC) in the absence of any multivalent ion condenses the polymer to toroidal structures at the midpoint (0.75 M NaCl) of the B to Z transition. Further addition of NaCl unfolds these toroids to rodlike structures, which show characteristic Z-form CD spectra.

Similarly, we have found a sequence of related secondary and tertiary structure changes as trivalent cations are added to a low salt solution of poly(dA-dT)·poly(dA-dT)[75]. This polymer undergoes a reversible conformational transition, to an undefined but definitely non-B structure as indicated by CD, in the presence of $Co(NH_3)_6^{3+}$ or spermidine in low salt (10 mM NaCl + 1 mM Na cacodylate). Under the salt conditions indicated, the CD transition begins with $Co(NH_3)_6^{3+}$ at about 70 µM and is complete by 150 µM; with spermidine, it begins at about 300 µM and is complete by 600 µM. Total intensity light scattering shows a marked increase at trivalent cation concentrations somewhat below those at which the CD transition begins. QLS measurement of the translational diffusion coefficient D_T shows that in the presence of $Co(NH_3)_6^{3+}$, the hydrodynamic radius R_h increases from 260 to 1450 Å over the concentration range 25-200 µM. With spermidine, R_h is 550±50 Å up to 200 µM, then increases rapidly. Values of R_h in this range are generally found for toroidal or other compact condensed forms of DNA. Such forms - toroidal, spheroidal, and rodlike structures - are observed in electron micrographs of poly(dA-dT)·poly(dA-dT) when the trivalent cation concentration is in the transition range. Above that range, extensive aggregation of the polymer chains is seen.

These strong correlations between aggregating tendency and non-B helix geometry, coupled with demonstrations that short sequences of alternating Pur-Pyr can adopt a Z-helical form even when inserted between regions that must remain in B form[76], make it worth considering that collapse of natural DNA is caused by association of short stretches of non-B DNA produced by interaction with metal ions or polyamines, or the dehydrating conditions produced by high alcohol concentrations. While CD or x-ray diffraction might not detect a few per cent of such structures, they should be discernable by Raman spectroscopy or by chemical or antibody probes. Experiments to test these ideas are currently underway in our laboratory.

ACKNOWLEDGMENTS

I am grateful to my collaborators for their parts in the work reviewed here. The research was supported by grants from NIH (GM 17855, GM 28093) and NSF (PCM 84-16305).

REFERENCES

1. A. W. Fulmer, J. A. Benbasat and V. A. Bloomfield, Biopolymers, 1981, 20, 1147.
2. M. G. Fried and V. A. Bloomfield, Biopolymers, 1984, 23, 2141.
3. R. W. Wilson and V. A. Bloomfield, Biochemistry, 1979, 18, 2192.
4. D. A. Knoll, M. G. Fried and V. A. Bloomfield, Proc 5th Conversation Biomol Stereodynam., Adenine Press, 1988, v2, 123.
5. K. S. Schmitz and M. Lu, Biopolymers, 1984, 23, 797-808.
6. J. M. Schurr and K. S. Schmitz, Ann. Rev. Phys. Chem., 1986, 37, 271.
7. K. Zero and B. R. Ware, J. Chem. Phys., 1984, 80, 1610.
8. L. Wang, M. M. Garner, M. T. Record and H. Yu, 1988, private communication.
9. S-C. Lin, W. I. Lee and J. M. Schurr, Biopolymers, 1978, 17, 1041.
10. D. Stigter, Biopolymers, 1979, 18, 3125.
11. L. Onsager, Ann. NY Acad. Sci., 1949, 51, 627.
12. A. J. Hurd, N. A. Clark, R. C. Mockler and W. J. O'Sullivan, Phys. Rev. A, 1982, 26, 2869.
13. A. J. Hurd, N. A. Clark, R. C. Mockler and W. J. O'Sullivan, J. Fluid Mech., 1985, 153, 401.
14. M. Drifford and J-P. Dalbiez, Biopolymers, 1985, 24, 1501.
15. M. Weissman, J. Chem. Phys., 1980, 72, 231.
16. P. N. Pusey and R. J. A. Tough, in *Dynamic Light Scattering Applications of Photon Correlation Spectroscopy*, 1985, R. Pecora, ed, Plenum, New York, 85.
17. A. Vrij, Chem. Phys. Lett., 1978, 53, 144-147; J. Chem. Phys., 1978, 69, 1742.
18. D. F. Evans, S. Mukherjee, D. J. Mitchell and B. W. Ninham, J. Colloid Interface Sci., 1983, 93, 184.
19. D. J. Mitchell, B. W. Ninham and D. F. Evans, J. Coll. Interface Sci., 1984, 101, 292.
20. K. S. Schmitz, M. Lu and J. Gauntt, J. Chem. Phys., 1983, 78, 5059.
21. K. S. Schmitz, M. Lu, N. Singh and D. J. Ramsay, Biopolymers, 1984, 23, 1637.
22. J. A. Benbasat and V. A. Bloomfield, Macromolecules, 1973, 6, 593.
23. R. L. Rill, P. R. Hillard and G. C. Levy, J. Biol. Chem., 1983, 258, 250.
24. T. E. Strzelecka and R. L. Rill, J. Am. Chem. Soc., 1987, 109, 4513.

25. M. Doi and S. F. Edwards, *Theory of Polymer Solutions,* 1986, Oxford Univ. Press.
26. V. A. Bloomfield, *Reversible Polymeric Gels and Related Systems,* 1987, in Russo P, ed. , ACS Symposium Series <u>350</u>, 199.
27. M. A. Krasnow and N. R. Cozzarelli, <u>J. Biol. Chem.</u>, 1982, <u>257</u>, 2687.
28. W. G. Nelson, K. J. Pienta, E. R. Barrack and D. S. Coffey, <u>Ann. Rev. Biophys. Biophys. Chem.</u>,1986, <u>15</u>, 457.
29. W. C. Earnshaw and S. R. Casjens, <u>Cell</u>, 1980, <u>21</u>, 319.
30. U. K. Laemmli, J. R. Paulson and V. Hitchins, <u>J. Supramol. Struct.</u>, 1974, <u>2</u>, 276.
31. L. S. Lerman, <u>PNAS</u>, 1971, <u>68</u>, 1886.
32. C. F. Jordan, L. S. Lerman and J. H. Venable Jr., <u>Nature New Biol.</u>, 1972, <u>232</u>, 67.
33. L. S. Lerman, <u>Cold Spring Harbor Symp. Quant. Biol.</u>, 1973, <u>38</u>, 59.
34. T. Maniatis, J. H. Venable Jr., L. S. Lerman, <u>J. Mol. Biol.</u>, 1974, <u>84</u>, 37.
35. W. C. Earnshaw, J. King, S. C. Harrison and F. A. Eiserling, <u>Cell</u>, 1978, <u>14</u>, 559.
36. L. W. Black, N. W. Newcomb, J. W. Boring and J. C. Brown, <u>PNAS</u>, 1985, <u>82</u>, 7960.
37. P. Serwer, <u>J. Mol. Biol.</u>, 1986, <u>190</u>, 509.
38. S. M. Klimenko, T. I. Tikchonenko and V. M. Andreev, <u>J. Mol. Biol.</u>, 1967, <u>23</u>, 523.
39. L. C. Gosule and J. A. Schellman, <u>Nature</u>, 1976, <u>259</u>, 333.
40. D. K. Chattoraj, L. C. Gosule and J. A. Schellman, <u>J. Mol. Biol.</u>, 1978, <u>121</u>, 327.
41. J. Widom and R. L. Baldwin, <u>J. Mol. Biol.</u>, 1980, <u>144</u>, 431.
42. G. C. Ruben and K. A. Marx, <u>Proc 43rd Meeting Electron Microscopy Soc Amer.</u>, 1985, 522.
43. A-Z. Li, P. Arscott and V. A. Bloomfield, <u>Biophy J.</u>, 1987, <u>51</u>, 499a.
44. A-Z. Li and V. A. Bloomfield, <u>Biophys J.</u>, 1986, <u>49</u>, 302a.
45. C. Post and B. H. Zimm, <u>Biopolymers</u>, 1982, <u>21</u>, 2123.
46. C. Post and B. H. Zimm, <u>Biopolymers</u>, 1982, <u>21</u>, 2139.
47. C. Tanford, <u>J. Chem. Phys.</u>, 1974, <u>78</u>, 2469.
48. S. C. Riemer and V. A. Bloomfield, <u>Biopolymers</u>, 1978, <u>17</u>, 785.
49. F. Oosawa, *Polyelectrolytes*, 1971, Marcel Dekker, New York, Ch. 9.
50. D. C. Rau, B. Lee and V. A. Parsegian, <u>Proc. Natl. Acad. Sci. U.S.A.</u>, 1984, <u>81</u>, 2621.
51. G. S. Manning, <u>Quart. Rev. Biophys.</u>, 1978, <u>11</u>, 179.

52. G. L. Eichhorn and Y. A. Shin, J. Am. Chem. Soc., 1968, 90, 7323.
53. S. Hiai, J. Mol. Biol., 1965, 11, 672.
54. G. L. Eichhorn and P. Clark, Proc. Natl. Acad. Sci. USA, 1965, 53, 586.
55. Y. A. Shin and G. L. Eichhorn, Biochemistry, 1968, 7, 1026.
56. A. P. Yurgaitis and Y. S. Lazurkin, Biopolymers, 1981, 20, 967.
57. V. L. Stevens and E. L. Duggan, J. Am. Chem. Soc., 1957, 79, 5703.
58. M. H. Zehfus and W. C. Johnson, Jr., Biopolymers, 1981, 20, 1589.
59. M. H. Zehfus and W. C. Johnson, Jr., Biopolymers, 1984, 23, 1269.
60. W. C. Johnson, Jr. and J. C. Girod, Biochim. Biophys. Acta, 1974, 353, 193.
61. S. M. Cheng and S. C. Mohr, FEBS Letters, 1974, 49, 37.
62. M. T. Record, C. F. Anderson and T. M. Lohman, Q. Rev. Biophys., 1978, 11, 103.
63. S. B. Zimmerman, Ann. Rev. Biochem., 1982, 51, 395.
64. K. M. Sullivan and D. M. Lilley, J. Mol. Biol., 1987, 193, 397.
65. A. Rich, A. Nordheim and A. H.-J. Wang, Ann. Rev. Biochem., 1984, 53, 791.
66. S. Devarjan and R. H. Shafer, Nucleic Acids Res., 1986, 14, 5099.
67. A. Woisard, W. Guschlbauer and G. V. Fazakerley, Nucleic Acids Res., 1986, 14, 3515.
68. M.-F. Hacques and C. Marion, Biopolymers, 1986, 25, 2281.
69. W. Zacharias, J. C. Martin and R. D. Wells, Biochemistry, 1983, 22, 2398.
70. H. Castleman and B. F. Erlanger, Cold Spring Harbor Symp. Quant. Biol., 1983, 47, 133.
71. Y. A. Shin and G. L. Eichhorn, Biopolymers, 1984, 23, 325.
72. J. H. van de Sande and T. M. Jovin, EMBO J., 1982, 1, 115.
73. J. H. van de Sande, L. P. McIntosh and T. M. Jovin, EMBO J., 1982, 1, 777.
74. T. J. Thomas and V. A. Bloomfield, Biochemistry, 1985, 24, 713.
75. T. J. Thomas and V. A. Bloomfield, Biopolymers, 1985, 24, 2185.
76. J. Klysik, S. Stirdivant, J. E. Larson, P. A. Hart and R. D. Wells, Nature, 1981, 290, 671.

13

15N NMR Spectroscopy of DNA in the Solid State

By S. J. Opella[1] and K. M. Morden[2]

[1]DEPARTMENT OF CHEMISTRY, UNIVERSITY OF PENNSYLVANIA, PHILADELPHIA, PENNSYLVANIA 19104, U.S.A.

[2]SMITH, KLINE AND FRENCH LABORATORIES, PHILADELPHIA, PENNSYLVANIA 19101, U.S.A.

1. INTRODUCTION

Although the genomes of all organisms take the form of compactly folded high molecular weight polymers of DNA (or RNA), nearly all of the structural information about DNA has come from studies of oligonucleotides or fibers. There is a real need to develop methods for describing the structure and dynamics of high molecular weight duplex DNA, since the structural properties of naturally occurring DNA, including its folding, must be strongly influenced by its being a polymer in a concentrated gel-like state rather than an oligonucleotide in a crystal or in solution.

Polymeric DNA has been studied by both x-ray fiber diffraction[1] and solid state NMR spectroscopy[2-10]. Fiber diffraction provides valuable structural information within its limitations of requiring highly oriented samples and yielding spatially averaged results. Solid-state NMR spectroscopy provides both structural and motional information and can be applied to both oriented and unoriented samples of DNA and can, in favorable cases, be used to study DNA in the presence and absence of proteins. These systems are difficult or impossible to study by x-ray diffraction, because of its requirements for single crystal samples, or solution NMR spectroscopy, because of its requirements for low molecular weight and rapidly reorienting samples. Overall, the development of solid-state NMR spectroscopy for the study of DNA and nucleoprotein complexes will extend the range of systems that can be characterized at atomic resolution, both in terms of the types of nucleic acids and the types of samples that can be used in the experiments. In addition, it will be possible to describe nucleic acid structure and dynamics in a truly integrated manner by solid-state NMR spectroscopy.

^{15}N NMR spectroscopy of uniformly ^{15}N labelled DNA is a powerful approach to describing the bases of DNA. Previously we have obtained high resolution ^{15}N NMR spectra of duplex DNA and the single stranded DNA in filamentous bacteriophages[6]. The resolved ^{15}N resonances in these spectra enable separated local field experiments to be used to measure the N-H bond lengths of the sites participating in hydrogen bonds between base pairs[7]. In this paper we demonstrate the applicability of ^{15}N spin-exchange spectroscopy for describing the three-dimensional structure of high molecular weight polymeric DNA. These results demonstrate that homonuclear ^{15}N dipole-dipole interactions provide a direct measure of distances that can be used to determine the structure of DNA. The combination of N-H and N-N distances available from heteronuclear separated local field and homonuclear spin-exchange experiments provides a powerful and general approach to structure determination of naturally occurring DNA in samples that can not be studied by other methods. In addition, several aspects of the spectroscopy important for the design and implementation of these experiments, including the field independence of the spectral resolution and the use of a resonance selection procedure, are demonstrated.

The chemical structures of the four DNA bases and their hydrogen bonding arrangements in base pairs are shown in Figure 1. The nitrogen sites in DNA are of chemical interest because of their participation in hydrogen bonds and are of spectroscopic interest as markers (on the bases) throughout the interior of the double helical structure. The DNA can be uniformly enriched biosynthetically to contain greater than 90% ^{15}N in all nitrogen sites. ^{15}N NMR spectra of uniformly labelled DNA have been obtained both in solution[11] and in the solid-state[6]. The isotropic chemical shift spectra display nearly complete resolution among the chemically distinct nitrogen sites in the bases of DNA. The high resolution solid-state ^{15}N NMR spectrum of high molecular weight DNA has been analyzed previously[6] and the results from experiments that measured heteronuclear dipolar couplings in order to determine the N-H bond lengths with high precision have been described[7].

Homonuclear spin-exchange occurs most rapidly between nuclei that are in close proximity. The longest distance that spin-exchange can occur over is not known,

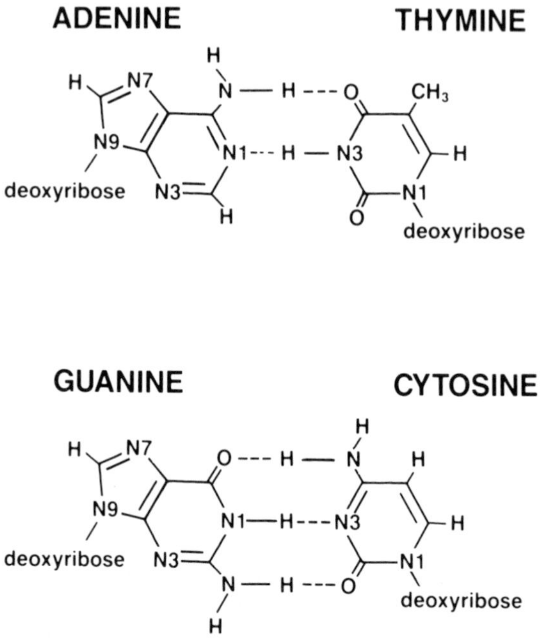

Figure 1: The nitrogen numbering scheme for the four bases commonly found in DNA. They are shown in the conventional Watson-Crick base pairs.

although inter-molecular exchange has been observed in the solid state[12]. In previous measurements on a protein, spin-exchange was observed between nitrogens separated by 2.6Å[13]. The nitrogens within any DNA base are separated by distances between 2.25 and 5.4Å. These distances will be
· referred to as intra-base distances. The nearest nitrogens across a base pair are 2.9Å apart and the furthest are 9Å. These distances will be referred to as intra-base-pair distances. In B-form DNA distances between nitrogens on neighboring bases in a stack vary between 3.4 and 9Å and will be referred to as inter-base-pair distances. Thus, there are potentially many distances between nitrogens in DNA which can be measured by nitrogen spin-exchange experiments.

2. MATERIAL AND METHODS

A single sample of DNA was used for all of the experimental measurements. This sample was obtained from E. coli infected with the filamentous bacteriophage fd. The DNA extracted from the bacteria was primarily in the form of the 6800 base pair duplex viral replicative intermediate. The bacteria were grown on a medium with $(^{15}NH_4)_2SO_4$ as the sole source of nitrogen. The DNA was purified by standard methods and precipitated with ethanol[6,14]. The DNA was then rehydrated to B-form (92% hydration) by equilibration with sodium tartrate in a hydration cell[15]. Hydration was monitored by gravimetric titration. The final sample weight was approximately 250 mg.

The NMR experiments were performed on a homebuilt double resonance spectrometer with a 3.5T magnet operating at 15.1 MHz for ^{15}N and on a modified JEOL GX-400 WB spectrometer operating at 40 MHz for ^{15}N. All spectra were obtained by cross-polarization with a 2 msec mix time, proton decoupling of 1.3 mT, and magic angle sample spinning of approximately 2.8 kHz. All of the ^{15}N NMR spectra are referenced to $(^{15}NH_4)_2SO_4$ having a chemical shift of zero. The two-dimensional spin-exchange spectra were obtained with a pulse sequence consisting of cross-polarization followed by a relatively long (2-6 sec) mix time where the proton decoupling is turned off and the nitrogen magnetization is aligned along the z axis by a 90° pulse[12,16-18]. The magnetization is then detected following another 90° pulse. The selective spin-exchange spectrum was obtained with a pulse sequence having delay intervals without proton irradiation both before and after the spin-exchange mix time that serve to select for magnetization and cross-peaks arising only from nitrogen sites without directly bonded protons[19].

3. RESULTS

Several one-dimensional ^{15}N NMR spectra of high molecular weight polymeric duplex DNA hydrated to be in the B-form are presented in Figure 2. These spectra were obtained with magic angle sample spinning and proton decoupling, therefore each resonance line is from a single type of nitrogen. There are fifteen chemically distinct nitrogens in DNA, as shown in Figure 1, and twelve resonances in the spectrum in Figure 2A. The nearly complete resolution among types of nitrogen presents favorable opportunities for high resolution spectroscopic studies.

Figures 2A and B compare ^{15}N NMR spectra obtained at ^{15}N resonance frequencies of 40 MHz and 15 MHz. The resonance linewidths in these two spectra scale

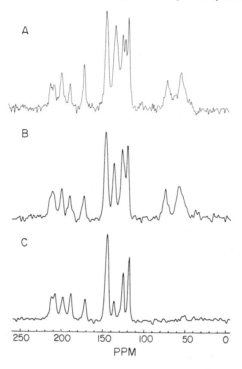

<u>Figure 2</u>: One-dimensional ^{15}N Cross Polarization Magic Angle Spinning spectra of hydrated DNA. A) Obtained at 40 MHz with a cross polarization mix time of 4 msec, a recycle delay of 10 sec and a spinning rate of 5.9 kHz.   The spectrum is the sum of 48 accumulations. B) Spectrum of the same sample obtained at 15 MHz with a cross polarization mix time of 2 msec, a recycle delay of 9 sec and spinning at 2.7 kHz.   The spectrum is the sum of 100 accumulations. C) Obtained under the same conditions given for spectrum B except using the SELPEN pulse sequence. The delay (without proton decoupling) before accumulation was 120 μsec. The spectrum is the sum of 500 accumulations. Chemical shifts are given relative to solid $(^{15}NH_4)_2SO_4$.

approximately with the resonance frequency. Therefore, it appears that the linewidth of each resonance is a consequence of chemical shift dispersion from chemically identical nitrogens which are influenced by local interactions, for example from neighboring bases. The 40 MHz NMR experiments have significantly higher sensitivity than the 15

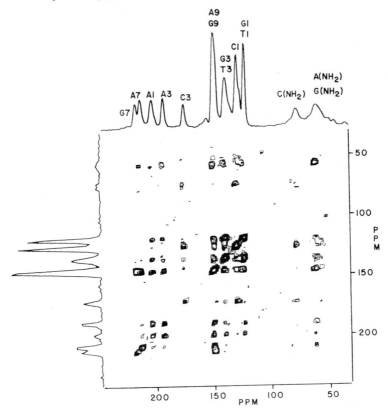

Figure 3: $^{15}N\text{-}^{15}N$ two-dimensional spin-exchange spectrum of hydrated DNA. The symmetrized contour plot is 256 x 256 points. Each t_1 value is the sum of 100 accumulations taken with a 2 msec cross polarization mix time and a 4 sec spin-exchange mix time.

MHz NMR experiments, as indicated by equivalent signal to noise ratios being obtained with less than half the number of scans at the higher frequency. In Figures 2B and C a comparison is made between a conventional high resolution solid-state NMR spectrum obtained with cross-polarization and a spectrum obtained using a selective pulse sequence. The selective pulse sequence, SELPEN, consists of a cross-polarization preparation period followed by a brief interval without proton irradiation during which the magnetization from the nitrogens with bonded protons is lost[20]. The experiment yields spectra with resonances only from nitrogens without bonded protons as shown in Figure 2C. This selection procedure, based on heteronuclear dipolar coupling, is useful in enhancing resolution and making assignments in both one- and two-dimensional NMR spectra. The assignments made in an earlier publication[6] are thus confirmed by this selective experiment at 40 MHz.

The two-dimensional homonuclear spin-exchange experiments work better and are simpler to interpret at the lower resonance frequency, despite the lower sensitivity, because there is greater spectral overlap among the resonances, enhancing the spin-exchange process, and it is possible to spin the sample fast enough to essentially eliminate spinning side bands, thus making it unnecessary to apply side-band suppression pulse sequences. The spin-exchange spectra presented in Figures 3-5 were obtained at 15 MHz.

Figure 3 contains the complete two-dimensional ^{15}N NMR spin-exchange spectrum of DNA. It is presented as a contour plot with the one-dimensional spectrum aligned along the top and the projection aligned along the side. The diagonal corresponds to the one dimensional chemical shift spectrum. The many cross- peaks in the spectrum display considerable variation in intensity. A region of this spectrum is shown expanded in Figures 4 and 5 under several different experimental conditions. The two-dimensional spectrum in Figure 3 and the expansion in Figures 4B and 5B are from the same data and were obtained with a mixing interval of 4 sec.

The spectra in Figures 4A-C compare the cross-peak intensities for mixing intervals of 2 sec, 4 sec, and 6 sec. This comparison shows that cross-peak intensities build up as a function of mix time. The peaks observable after only a 2 sec mix time can all be attributed to spin-exchange between nitrogens that are close in space and on the same base. With a 4 sec mix time the number of cross-peaks has increased over the number observed in the spectrum obtained with a 2 sec mix time and several of the new cross- peaks can be identified as due to intra-base-pair nitrogen-nitrogen interactions. The 6 sec interval yields additional cross-peaks compared to the 4 sec mix time.

Figure 5 contains a different comparison for this same expanded spectral region. Both of the two-dimensional spectra were obtained with the same 4 sec mix time. However, the spectrum in Figure 5A contains only

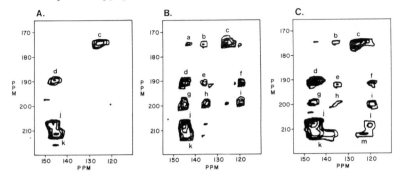

Figure 4: Expanded region of ^{15}N-^{15}N spin-exchange spectra obtained with different exchange mixing times, τ_m. A) τ_m = 2 sec, B) τ_m = 4 sec, C) τ_m = 6 sec. The full spectrum for figure B is shown in Figure 3. The assignments of the cross peaks in these spectra are as follows: a) G(N3) to G(N9), b) C(N3) to G(N3), c) C(N1) to C(N3), d) A(N3) to A(N9), e) A(N3) to T(N3), f) T(N1) to A(N3), g) A(N1) to A(N9), h) A(N1) to T(N3), i) A(N1) to T(N1), j) A(N7) to A(N9), k) G(N7) to G(N9), l) T(N1) to A(N7) and m) G(N1) to G(N7).

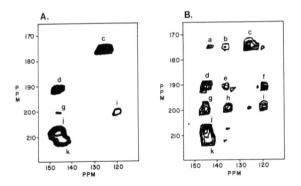

Figure 5: Expanded region of two-dimensional ^{15}N-^{15}N spin-exchange experiments. A) Selective experiment using SELEX pulse sequence containing a delay before the spin-exchange mix time and before accumulation. Delay time (without proton decoupling) was 120 μsec. B) Standard spin-exchange experiment, with a exchange mix time of 4 seconds. The full spectrum is shown in Figure 3.

cross-peaks resulting from interactions between nitrogens without bonded protons. This is because the heteronuclear dipolar interaction was used to select for the resonances from nitrogens without bonded protons in a manner analogous to the one-dimensional spectrum shown in Figure 2C[19]. Additional relaxation can occur when selecting for the resonances from the nitrogens without bonded protons in the spin-exchange experiment, since delays without proton irradiation must precede and follow the mixing interval. The selection procedure was successful, since the cross-peaks present in Figure 5B and missing in Figure 5A associated with resonances where at least one nitrogen site has a bonded proton.

4. DISCUSSION

The ^{15}N NMR spectrum shown in Figure 2B differentiates among chemically distinct nitrogens. However, there is no evidence for differentiation resulting from interactions with neighboring bases in the DNA. This is not surprising considering all of the possible combinations of neighbors and nearest neighbors present in DNA with 6800 base pairs. Therefore, we have made cross-peak assignments on the basis of the most probable interaction. Intra-base and intra-base-pair interactions are the most probable due to the abundance of these interactions and their having, in general, shorter internuclear distances than inter-base-pair interactions. Thus, in Figures 4 and 5 the cross-peaks are assigned to interactions that are across a base pair and have not been assigned to any of the potential inter-base-pair interactions.

Some examples of distances between nitrogens in a three base pair stack in an idealized duplex of B form DNA are shown in Figure 6.

The nitrogens within a base pair are separated by 2.25 to 5.4 Å. Cross-peaks due to spin-exchange between many of these nitrogens are readily observable, even for relatively short mix times, as shown in Figure 4A. The nearest nitrogens across a base pair are separated by 2.9 Å. These cross-peaks are present in the spectra, as shown in Figure 4B. In B form DNA the bases stack vertically with a separation of about 3.4 Å and a winding angle of 36° per base pair. There is a considerable range of distances between nitrogens on bases above and below, extending from 3.4 to 9Å.

In Figure 3, the results of the two-dimensional

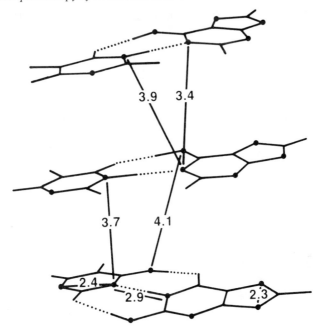

<u>Figure 6</u>: Base pairs stacked in a standard B-form helical conformation, containing a rise of 3.38 Å and a winding angle of 36°. Coordinates were generated according to Arnott and Hukins (1972). The three base pair stack is dC-A-A + dT-T-G. The filled circles represent nitrogens.

spin-exchange experiments are shown. In this spectrum the cross-peaks indicate the interactions between nitrogens which are close to each other. The interaction is influenced by the length of the spin-exchange mixing time and the relaxation rates of the different nitrogens, as well as inter-nuclear distance. Various experimental parameters strongly affect the intensity and number of cross-peaks that are observed.

The results of increasing the spin-exchange mix time are shown in Figure 4. The shortest mix time (Figure 4a) shows four strong cross-peaks which are all assigned to intra-base interactions. They represent interactions between C(N1) and C(N3), A(N3) and A(N9), A(N7) and A(N9), and G(N7) and G(N9). As the mix time is increased, more cross-peaks are observed and it is possible to attribute cross-peaks to both intra-base-pair and intra-base interactions. For instance, the interaction between A(N1) and T(N3) is shown as cross-peak h. As the mix time is increased to 6 sec (Figure 4C) several new cross-peaks appear which can be attributed to long range interactions. One example of this is cross-peak I which represents the interaction between A(N7) and T(N1), a distance of 8.32Å. Some of the cross-peaks, for example e and h, decrease in intensity compared to those observed at the shorter mix time of 4 sec. This loss in intensity is probably due to a shorter relaxation time for one of the nuclei involved in the interaction. Cross-peaks e and h represent interactions between T(N3) and another nitrogen. The N3 position of thymidine has a directly bonded proton and therefore, may have a faster relaxation rate. Thus we can see, in a qualitative manner, by combining data from a variety of mixing times and knowing the relative relaxation rates of these nitrogens that we can treat these data in a similar manner to those from solution NOESY experiments.

Because the spectra of biological molecules can become quite complex, it is essential to develop methods for spectral simplification and resonance assignment. One such method is demonstrated in Figure 5. In this Figure, results of two different 2D experiments are compared. Figure 5B shows the results of the standard 2D spin-exchange experiment with a 4 sec mix time. The spectrum shown in Figure 5A is the result of a 2D spin-exchange experiment which selectively observes only resonances from nitrogens without bonded hydrogens. This pulse sequence is called SELEX to denote the selective nature of the spin-exchange which is observed [19]. There is a decrease in the number of cross-peaks present in the spectrum. The cross-peaks that remain due to interactions between C(N1) and C(N3), A(N3) and A(N9), A(N1) and A(N9), A(N7) and A(N9), G(N7) and G(N9), and A(N1) and T(N1). All of these interactions are between non-proton bearing nitrogens. Missing from the spectrum in Figure 5A are peaks e and h; both involve interactions with T(N3) which has a directly bonded proton. During the additional delays in the SELEX pulse sequence, relaxation occurs so that cross-peaks between nitrogens with short relaxation times may not be observed and there may be an overall decrease in the signal to noise ratio. Other cross-peaks that are missing in Figure 5A are a, b, and f. These are the peaks from Figure 5B which represent the longest distance interactions, all of them being over 5 Å with weaker dipolar interactions.

One of the major problems with the current technology in solution NMR spectroscopy is the limitation in the size of the molecules that can be studied. This limitation is due in part to increased complexity of the spectrum, but perhaps more importantly to an increase in the correlation time of the molecule making it impossible to observe resonances in solution. For nucleic acids, large polymers must be broken down into fragments on the order of fifty bases in length in order for reasonable solution spectra to be obtained. Solution studies, until recently, have also focused on ^{1}H NMR with very few studies using either ^{13}C or ^{15}N. One study has recently reported the observation of ^{15}N resonances from DNA but was again restricted to DNA only 50 base pairs in length [11]. As demonstrated previously [6] and in this paper it is possible to obtain high resolution ^{15}N NMR spectra of a 6800 base pair fragment of DNA using solid-state NMR techniques. Not only do the one-dimensional spectra demonstrate nearly complete resolution among types of nitrogens, but also two-dimensional spin-exchange spectra have the potential for providing resonance assignments and providing structural information about DNA. The observation of ^{15}N magnetization has the advantage that there are several ^{15}N-^{15}N distances which are fixed by the chemical structure of the bases. For the analogous case of ^{1}H solution NMR on nucleic acids there is only one fixed distance (C(H5) to C(H6)) with which to calibrate the NOEs observed within the bases.

5. ACKNOWLEDGMENTS

We thank Dr. K. M. Valentine for providing the labeled DNA sample, Dr. J. A. DiVerdi for help with the instrumentation and Dr. T. A. Cross for helpful discussions. This research was supported, in part, by Grant GM-24266 from the N.I.H.

6. FOOTNOTES

≠ Present address: Department of Biochemistry, Louisiana State University, Baton Rouge, Louisiana 70803, USA.

* Address correspondence to this author at the University of Pennsylvania.

7. REFERENCES

1) Watson, J. and Crick, F.H.C., <u>Nature</u>, 1953, <u>171</u>, 736.

2) Shindo, H., Wooten, J. B., Pheiffer, B. H., & Zimmerman, S. B., Biochemistry, 1980, 19, 518.

3) Opella, S. J., Wise, W. B., & DiVerdi, J. A., Biochemistry, 1981, 20, 284.

4) Nall, B.T., Rothwell, W.P., Waugh, J.S., and Rupprecht, A., Biochemistry, 1981, 19, 518.

5) DiVerdi, J. A., & Opella, S. J., J. Mol. Biol., 1981, 149, 307.

6) Cross, T. A., DiVerdi, J. A., & Opella, S. J., J. Am. Chem. Soc., 1982a, 104, 1759.

7) DiVerdi, J. A., & Opella, S. J., J. Am. Chem. Soc., 1982, 104, 1761.

8) Mai, M. T., Wemmer, D. E., & Jardetsky, O., J. Am. Chem. Soc., 1983, 105, 7149.

9) Fujiwara, T., & Shindo, H., Biochemistry, 1985, 24, 896.

10) Vold, R.R., Brandes, R., Tsang, P., Kearns, P.R., Vold, R.L., & Rupprecht, A., J. Amer. Chem. Soc., 1986, 108, 302-303.

11) James, T. L., James, J. L., & Lapidot, A., J. Am. Chem. Soc., 1981, 103, 6748.

12) Caravatti, P., Deli, J. A., Bodenhausen, G., & Ernst, R. R., J. Am. Chem. Soc., 1982, 104, 5506.

13) Cross, T. A., Frey, M. H., & Opella, S. J., J. Am. Chem. Soc., 1982b, 104, 7471.

14) Smith, M. G., Meth. Enzymol., 1966, 12A, 545.

15) O'Brien, F. E. M., J. Sci. Instrum., 1948, 25, 73.

16) Suter, D., & Ernst, R. R., Phys. Rev. B, 1982, 25, 6038.

17) Szeverenyi, N. M., Sullivan, M. J., & Maciel, G. E., J. Magn. Reson., 1982, 47, 462.

18) Szeverenyi, N. M., Max, A., & Maciel, G. E., J. Am. Chem. Soc., 1983, 105, 2579.

19) Morden, K. M., & Opella, S. J., J. Magn. Reson., 1986, 70, 476.

20) Opella, S. J., & Frey, M. H., <u>J. Am. Chem. Soc.</u>, 1979, <u>101</u>, 5854.

21) Arnott, S., & Hukins, D. W. L., <u>Biochem. Biophys. Res. Commun.</u>, 1972, <u>47</u>, 1504.

Part III: Dynamics of Glycoconjugates and Membranes

14

Conformational Equilibria and Dynamics of Small Carbohydrates: Anomers, Tautomers, and Rotamers

By F. Franks

PAFRA LTD., BIOPRESERVATION DIVISION, 150, CAMBRIDGE SCIENCE PARK, CAMBRIDGE CB4 4GG, U.K.

1 INTRODUCTION

Polyhydroxy compounds (PHCs), along with amino acids, lipids and the nucleotide bases form the chemical building blocks of living matter. Their involvement as monomers, oligomers and polymers ranges from such well-understood attributes as sources of metabolic energy (starch) and load-supporting structures (cellulose), to rather more obscure functions, such as recognition (immune responses) and freeze avoidance (Antarctic fish antifreeze glycoproteins). A vast literature describes the chemistry of sugars, their derivatives and polymers. Most of it is devoted to organic chemistry and biochemistry: synthesis, derivatization and structures. Recently interest has shifted to the involvement of oligosaccharide residues O- or N-linked to proteins in biological recognition phenomena. Attention of physical chemists is now being focussed on the nature of weak, non-covalent, solvent-mediated interactions which determine the conformations and intramolecular rearrangements of monomeric and oligomeric PHCs.

This article discusses three examples of such stereochemical problems and aims to show how the combination of experimental (n.m.r.) and theoretical approaches (computer simulation) can be brought to bear on their resolution.

2 SUGAR EQUILIBRIA IN SOLUTION

The physico-chemical behaviour of simple sugars in solution is obscured by their chemical heterogeneity. Under equilibrium conditions the solutions contain mixtures of two or more anomers, conformers and/or tautomers, the proportions of which vary with temperature, concentration and the nature of the solvent in a manner which has been well documented but is not as yet fully

213

understood.(1-3) Attempts to calculate, and even to predict the equilibrium compositions of sugar solutions, have hardly been successful.

Monosaccharides are somewhat arbitrarily divided into two groups depending on whether they undergo simple or complex mutarotation.(1) The former process exhibits simple first-order kinetics and involves only two anomeric species, whereas the latter conversion involves several tautomeric species. The type of mutarotation observed depends on the relative rates of ring opening and closing and of pyranose/furanose conversions. The relationships between sugar stereochemistry, type of mutarotation and equilibrium composition have been formalised into a set of "rules" which are, however, of limited value because they are based either on crystal structures or on practical experience with aqueous solutions. Only scattered information exists about the influence of solvent and temperature on anomeric and tautomeric equilibria.(2) It was therefore of interest to compare two monosaccharides of which one (glucose) exhibits simple mutarotation and the other (ribose) exists as a complex equilibrium mixture. Dimethyl sulphoxide (DMSO) was chosen as the nonaqueous solvent, and ^{1}H n.m.r. as the experimental method.

Summaries of the results are shown in Figs. 1 and 2.(4) Glucose exists as a simple mixture of α- and β-pyranose (C1) forms in both solvents, although the temperature dependences of the α/β ratios differ significantly. Ribose exists as a complex mixture of C1 and 1C pyranose and furanose tautomers, in addition to the anomeric forms, i.e. six distinguishable species in all. The 1C and C1 α-pyranose forms could not be resolved because their H(1) and H(2) protons give rise to identical ^{3}J coupling constants.

The results indicate that β-pyranose-C1 is the favoured form, particularly at low temperatures and in aqueous solution, a fact which is at odds with calculated α/β ratios but in harmony with a proposed model for stereospecific hydration, made first by Kabayama and Patterson (5) and backed up by direct hydration studies based on n.m.r. and dielectric relaxation measurements on a series of sugars, e.g. (6). The model is based on the spatial compatibility of the -OH topology in water with that of equatorial -OH groups on pyranose sugars. In the sugars the spacings between oxygen atoms linked to next-nearest carbon neighbours are 0.485 nm which is also the distance between next-nearest neighbour oxygen atoms in liquid water.(7) Figure 3 illustrates how the glucose anomers could interact with an idealized, unperturbed water lattice, without causing any strain

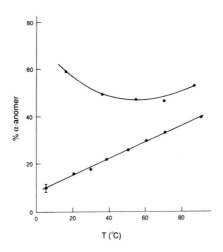

<u>Figure 1</u>   The equilibrium composition of glucose in D$_2$O (◆) and DMSO-d$_6$ (●) as function of temperature. After ref. 4.

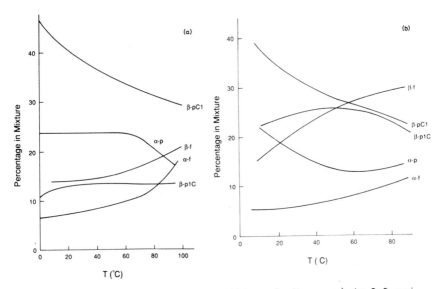

<u>Figure 2</u>   The equilibrium composition of ribose: a) in D$_2$O and b) in DMSO-d$_6$ as function of temperature; p = pyranose, f = furanose. After ref. 4.

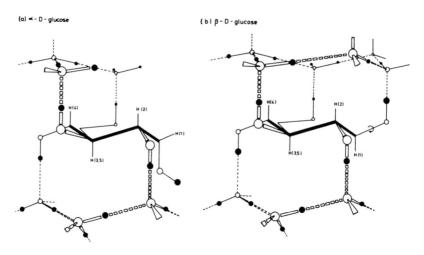

Figure 3 D-glucose hydrogen bonded into a hypothetical tetrahedral water structure, according to the specific hydration model. Water molecules below and above the plane of the sugar ring are shown: a) α-form, b) ß-form. The pyranose ring is indicated by the prominent filled-in line. Oxygen and hydrogen atoms are represented by open and filled-in circles; covalent and hydrogen bonds by solid and broken lines. The hydroxymethyl protons [H(6)] are omitted for sake of clarity. Reproduced, with permission, from ref. 6.

in the hydrogen bonding network. In a mixture of several isomeric species with similar free energies, any hydration contribution to the free energy would favour that conformer with the maximum number of equatorial -OH groups, and this is indeed observed.

Direct studies of PHC hydration are beset by many problems, arising mainly from the fact that the life times of hydration "structures" in solution are generally shorter than the time scales of most experimental methods. The incomplete realization of this problem has led to much confusion and many misunderstandings regarding the existence and nature of so-called "bound" water. Computer simulation, provided that the necessary care is taken in the formulation of potential functions, can be of great help in comparing hydration geometries and exchange rates, at least to a first approximation.

The only available Molecular Dynamics (MD) results of glucose in aqueous solution demonstrate the effect of water on the dynamics of the sugar.(8) The solvent acts so as to increase the structural fluctuations undergone by the sugar molecule. The radial distribution functions of water in the hydration sphere (defined as nearest neighbour molecules) show up clear differences between hydration structures surrounding $-CH_2OH$, axial and equatorial -OH groups. Despite the well-defined and localised hydrogen bonding between sugar -OH groups and water, all first hydration shell water molecules exchange rapidly with bulk water. In terms of the dynamics, therefore, water cannot be described as "bound" by the sugar.

It is thus established that the solvent has a measurable influence on the intramolecular transformations of PHC molecules and that such solvent effects are also sensitive to the stereochemical details (-OH group topology) of the PHC molecule. It is also clear that the solvation effects of different non-aqueous solvents are similar, but they differ substantially from those observed in aqueous solutions, with measurable differences even recorded between H_2O and D_2O!(9)

3 SOLVENT EFFECTS ON SUGAR ALCOHOL CONFORMATION

In view of the unavoidable chemical heterogeneity of sugars in solution, it might be thought that such complexities could not exist in acyclic molecules which, chemically, bear a similarity to sugars. The availability of stable high-field magnets has made possible the complete resolution of the [1]H n.m.r. spectra of sugar alcohols. The comparison of the diastereoisomers ribitol, xylitol and arabinitol in aqueous and nonaqueous solution illustrates that the stereochemical problems which beset sugars still exist, although in a more subtle way.

It has been stated (by crystallographers) that, in solution, the alditol molecules retain the conformations that exist in the crystal,(11) *i.e.* planar zig-zag (arabinitol) or sickle-shaped (xylitol, ribitol), depending on the molecular -OH topology.

The ^{1}H spectrum of arabinitol in pyridine-d5, measured at 620 MHz, is shown in Fig. 4.(10). Similar spectra also exist for the other two alditols, also in D_2O solution. The measured [3]J coupling constants indicate rapidly interconverting mixtures of *trans* and *gauche* conformers. With the aid of the known crystal structures and the appropriate Karplus equation, the [3]J values for the molecules in the crystalline state have been

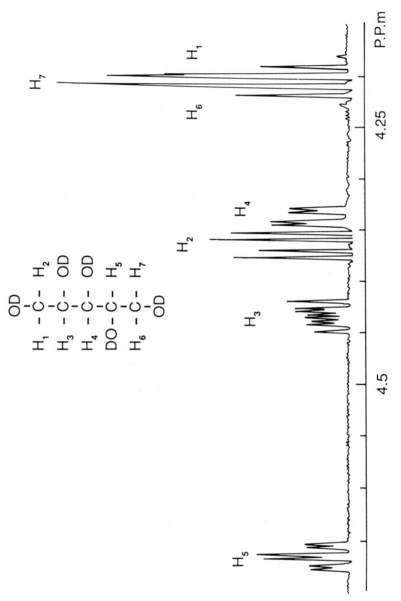

Figure 4 620 MHz ^{1}H n.m.r. spectrum of arabinitol in pyridine-d$_5$ at 25°C, with chemical shift assignments. From ref. 10.

calculated, as shown in Fig. 5 which also includes the measured solution ^3J values plotted against the corresponding dihedral angles *in the crystal*. The fractions of each conformer in the *trans* conformation in solution are summarized in Table 1. At a high enough temperature one would expect the molecules to undergo free rotation about each bond and the entries in Table 1 would each be 33%. The average departure from this "free rotation" value for each isomer is shown in Table 2.

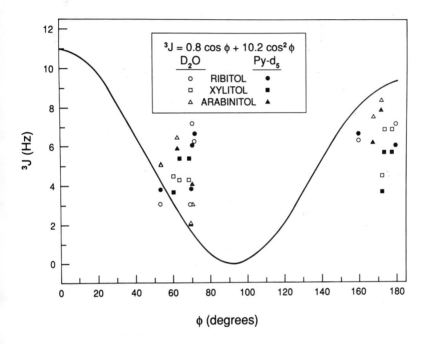

<u>Figure 5</u>   Plot of the Karplus equation <u>vs.</u> the H-H dihedral angle in pentitol crystals. The points give the experimental ^3J values: o, ribitol; xylitol; arabinitol. Open symbols, D_2O; filled symbols, $[^2H_5]$pyridine. $^3J = 0.8\cos\phi + 10.2\cos^2\phi$.

Two effects can be distinguished: In pyridine the alditols have a greater degree of rotational freedom than in D_2O, presumably due to decreased solvation effects. Of greater importance is the fact that, in D_2O, xylitol shows significantly freer rotation than either of the other isomers. Whereas the crystal structures suggest a similarity in conformation between

Table 1   Percentage   H–H *trans*-rotamers in the pentitols in
solution compared to the conformations in the crystal

	ribitol			xylitol			arabinitol		
	crystal	D_2O	Py	crystal	D_2O	Py	crystal	D_2O	Py
H_1–H_3	g	0	13	g	20	38	g	32	49
H_2–H_3	g	65	47	t	60	42	t	71	49
H_3–H_4	t	51	57	g	23	10	g	14	14
H_4–H_5	g	51	57	t	23	10	t	85	77
H_5–H_6	g	0	13	g	20	38	g	0	18
H_5–H_7	t	65	47	t	60	42	g	54	46

Table 2   Percentage departure from free rotation

	ribitol	xylitol	arabinitol
D_2O	28	17	30
Py–d_5	19	12	23

ribitol and xylitol, thermodynamic solution properties indicate a similarity between arabinitol and ribitol and unique behaviour for xylitol.(10) The results in Table 2 therefore support the view that the thermodynamic properties are a reflection of the conformation in solution and are not related to the crystal structure.

Although no theoretical or simulation data are available for pentitols, Grigera has performed MD simulations on the two diastereoisomeric hexitols mannitol and sorbitol which differ only in the position of the -OH group on C(4).(12) For the *in vacuo* simulation, no difference could be observed in the end-to-end distances or the radii of gyration (0.54 nm) of the two molecules. However, in aqueous solution ("computer water"), the radii of gyration decreased to 0.417 nm for mannitol and 0.372 nm for sorbitol, irrespective of the starting configuration, i.e. planar zig-zag or bent carbon chain. These values were peculiar to the aqueous solvent and differed markedly from those obtained with a Lennard-Jones solvent or an argon-like solvent in which the Lennard-Jones parameters had been chosen to match those of the water oxygen atom. The hydration dynamics are also of interest: the residence times of water molecules in the alditol hydration shell (i.e. within 1.2 molecular diameters), normalized to the residence time of a water molecule next to another water molecule, are 0.80 for mannitol and 0.39 for sorbitol. If one relates, as is commonly (but incorrectly) done, structural implications to dynamic data, then the "structure breaker" classification of both alditols is justified, with sorbitol producing a larger disruption in water structure. Here again, emphasis must be on the subtleties in the hydration interactions, feeding back into specific changes in the conformations of chemically almost identical molecules.

4 OLIGOSACCHARIDE CONFIGURATION

In disaccharides and higher oligomers conformational complexity is due more to the inherent flexibility of such molecules than to the coexistence of anomeric and tautomeric species. The configurational degrees of freedom are expressed in terms of the torsional angles which measure the rotation of two sugar residues about the glycosidic linkage. Increasing attention is being paid to oligosaccharide conformations in the belief that structural detail has important implications for the biological functions of such sugar clusters, e.g. in the blood group determinants. Although pyranose residues usually adopt the same conformations which exist in the monomeric sugars, this is probably not true for furanose rings, because of the small energy differences between several alternative ring conformers.

Until the development of very high resolution n.m.r. methods, torsional angles were estimated from optical rotation measurements. Despite the interpretative problems inherent in such measurements, Rees and his colleagues used such methods to good effect,(13) by expressing the optical rotatory power of a disaccharide as a sum of contributions from the two sugar residues and from the glycosidic linkage. The latter contribution is related to the particular values adopted by the torsional angles. A comparison of the experimentally determined linkage rotations for the diglucoses trehalose, cellobiose and methyl-α-D-maltoside in water, DMSO and dioxan with those calculated from crystal structure data and from conformational energy calculations on isolated molecules clearly shows up major differences between crystal, vacuum and solution conformations. The calculated *in vacuo* conformations agree with the experimental values for non-aqueous solvents, but the disaccharide conformations in aqueous solution (and also the temperature dependence of the linkage rotation) differ markedly from those calculated from crystal structures and those experimentally determined for organic solvents. Where crystal and solution data diverge, this may be due to inter- and/or intramolecular hydrogen bonds in the crystal which are replaced in solution by sugar-water bonds.

Nowadays more direct methods for establishing rotamer conformations exist in the form of n.m.r. ^3J coupling constants and nuclear Overhauser enhancement spectroscopy (NOESY). Of the disaccharides, sucrose has received most attention. There is as yet no universal agreement about a preferred solution conformation or the degree of rotational freedom about the glycosidic bond. It has been claimed that in concentrated solution two intramolecular hydrogen bonds are formed and the molecule is forced to take up the conformation which exists in the crystal;(14) in dilute solution these bonds are believed to be absent. In contrast, Bock and Lemieux claim that the sucrose molecule, even in dilute solution, is characterized by its rigidity, due to an intermolecular hydrogen bond between O-1(f) and O-2(g) (f = fructose, g = glucose).(15) Their published Ramachandran energy map, calculated on the basis of HSEA (hard-sphere-exoanomeric-effect) potential functions, exhibits two deep energy wells differing by no more than 10 kJ mol^{-1}, and corresponding to two possible conformational states both of which exist in the crystal.

Despite the undoubted thoroughness of their n.m.r. experiments, the conclusions reached by Bock and Lemieux cannot be accepted as the last word. The persistence of intramolecular hydrogen bonds giving rise to a rigid molecule in water solvent is unlikely. Also the claim that the conformation of sucrose is insensitive to the nature of the solvent runs counter to earlier experience with other disaccharides and needs further investigation. As to the calculated energy map, the methods used, i.e. hard-sphere vacuum potentials which are repulsive in character, cannot take into account any of the subtle stereospecific and solvation effects, already referred to above. The explicit inclusion of a contribution to the force field arising from the anomeric effect must also cast doubt on the credibility of such calculations; such contribution to the total potential function cannot be anything better than an adjustable quantity.

In those instances where thorough studies have been performed, the molecular conformations in the crystal state do not coincide with the solution conformations. It is therefore surprising that in the isolated sucrose molecule, deep potential wells should exist, corresponding to those in the crystal in which the molecule is extensively hydrogen bonded to several other sucrose molecules.

More complex saccharide structures are also receiving increasing attention by theoreticians and experimentalists alike. Particular interest centres on the N-asparagine-linked clusters of the blood group glycoproteins. A typical example is provided by the blood group A tetrasaccharide

$$Fuc(\alpha1{\to}2)$$
$$\searrow$$
$$\qquad\qquad Gal(\beta1{\to}3)\text{-}GalNAc\text{-}ol$$
$$GalNAc(\alpha1{\to}3)\nearrow$$

which has been studied by a combination of n.m.r. and theoretical methods.(16) Model calculations were performed on the two disaccharides Fuc(α1->2)Galβ-OMe and GalNAc(α1->3)Galβ-OMe. The calculations which included three different potential energy functions, gave several minimum energy conformations. NOESY measurements, on the other hand, suggest a rather restricted area in the conformational energy map compatible with experimental observation. The possibility that the observed spectra could reflect an average of several conformations was ruled out as "...that seems quite unlikely", without further discussion. However, the interconversion rate between conformers might be fast on the n.m.r. time scale. Although the heights of the energy barrier are not given, they can be estimated from the contour plots to range from 70 MHz to 1GHz.

When the energy data are combined with those for the other model disaccharide, the conformation of the trisaccharide can be deduced. Here the different model potentials yield different low-energy conformations; the authors conclude that "...it is difficult to judge what is the conformation, if indeed there is a unique fixed conformation."

Many questions remain: which energy contributions are important in determining molecular shape; is there a unique conformation or are the results diagnostic of an averaging over several rotamer conformations; does saccharide function depend on unique "native" configurations and are they subject to cooperative transitions, analogous to those of proteins?

From the probable biological significance of multiple conformations of N-linked oligosaccharides it can be argued that the appropriate conformations are selected when the oligosaccharides are bound to proteins, either covalently or as ligands. It has been suggested that the stabilization of individual conformational subsets at different glycosylation sites of glycoproteins may explain the differential site-specific activities of glycosyl transferases at the various glycosylation sites.(17)

The biological relevance of secondary oligosaccharide structures is difficult to assess, and no correlation appears to exist between the primary structural type and the secondary structure, although it is likely that minor primary sequence changes can produce significant changes in shape (<u>viz.</u> protein mutants) and/or biological activity.

5 CONCLUSIONS

PHCs have for long been the Cinderellas of physical chemistry. Conformational effects, modulated by stereospecific solvation interactions and inter- and/or intramolecular hydrogen bonding are subtle but may have important practical consequences. Simplistic models, based on such concepts as hard sphere potentials or "bound" water, and the indiscriminate use of the anomeric effect as an adjustable parameter to fit experimental data, have led to erroneous conclusions regarding the "simple" behaviour of aqueous solutions. By means of a few typical examples we have tried to show that the predictive value of such models is limited and that, both structurally and dynamically, PHC solutions of real challenges to experimentalists and theoreticians alike. With the increasing importance of glycobiology,(18) it is to be hoped that the combination of advanced experimental, theoretical and computational techniques will be applied to resolve the complexities of PHCs in solution and *in vivo*.

REFERENCES

1. R.S. Shallenberger, *Advanced Sugar Chemistry*, Ellis Horwood, Chichester, 1982.
2. S.Y. Angyal, *Adv. Carbohydrate Chem. Biochem.* 42, 15(1984).
3. F. Franks, *Pure & Appl. Chem.*, 59, 1189(1987).
4. F. Franks, P.J. Lillford and G. Robinson, *J. Chem. Soc., Faraday Trans. I* in press.
5. M.A. Kabayama and D. Patterson, *Can. J. Chem.* 36, 658(1958).
6. A. Suggett, *J. Solution Chem.* 5, 33(1976).
7. J.C. Dore, *Water Sci. Rev.* 1, 3(1985).
8. J.W. Brady, *Carbohydrate Res.* 165, 306(1987).
9. A. Suggett and A.H. Clark, *J. Solution Chem.* 5, 1(1976).
10. F. Franks, R.L. Kay and J. Dadok, *J. Chem. Soc., Faraday Trans. I*, 84, 2595(1988).
11. H.S. Kim and G.A. Jeffrey, *Acta Crystallogr., Sect. B*, 25, 2607(1969).
12. R. Grigera, *J. Chem. Soc., Faraday Trans. I*, 84, 2603(1988).
13. D.A. Rees and D. Thom, *J. Chem. Soc. Perkin Trans. II*, 191(1977).
14. M. Mathlouti and C. Luu, *Carbohydrate Res.* 81, 203(1980).
15. K. Bock and R.U. Lemieux, *Carbohydrate Res.* 100, 63(1982).
16. C.A. Bush, Z.-Y. Yan and B.N. Narasinga Rao, *J. Amer. Chem. Soc.*, 108, 6188(1986).
17. A. Marsden, B. Robson and J.S. Thomson, *J. Chem. Soc. Faraday Trans. I*, 84, 2519(1988).
18. T.W. Radedmacher, R.B. Parekh and R.A. Dwek, *Ann. Rev. Biochem.*. 57, 785(1988).

15

Hydrodynamic Modelling of Complement

By S. J. Perkins

DEPARTMENT OF BIOCHEMISTRY AND CHEMISTRY, ROYAL FREE HOSPITAL SCHOOL OF MEDICINE, ROWLAND HILL STREET, LONDON NW3 2PF, U.K.

1 INTRODUCTION

The complement cascade of immune defence is composed of some 20 glycoproteins in plasma which are activated sequentially by limited proteolysis in response to signals associated with the entry of foreign material into the circulation.[1] The classical pathway of activation involves the recognition of immune complexes of IgG and IgM antibodies by the component C1q, and the subsequent activation of the components C1r, C1s, C4, C2 and C3. This activation is regulated by the components C1 inh, C4BP and Factor I. The alternative pathway of activation involves the activation of C3, followed by the recruitment of Factor B and the Factor D-dependent generation of the complex of activated C3 and Factor B, namely C3b.Bb. This enzyme amplifies C3 activation, and is regulated by Factor H, Factor I and properdin. Activated C3 is then able to activate C5. Complement generates a plethora of biological activities directed against these foreign materials in order to eliminate them. This is most directly manifested by the interactions between the components C5, C6, C7, C8 and C9 to lead to the assembly of the membrane attack complex. This disrupts membranes by inserting itself into them to form open pores.

Low-resolution structural studies have been performed on many of the complement components by a variety of means, such as electron microscopy, X-ray and neutron solution scattering, and ultracentrifugation. Electron microscopy is useful in being able to visualise images directly, but is prone to artifacts caused by the need to work in vacuo, beam damage, and the use of stains. Solution scattering[2,3] is useful in that the structures are measured in conditions that are close to physiological and that it is a multiparameter technique offering the ability (a) to derive the external dimensions of the molecule, (b) to obtain information on the internal arrangement of distinct chemical entities within the macromolecule, and (c) to calculate molecular weights. Hydrodynamic methods yield the sedimentation or diffusion coefficient $s^{\circ}_{20,w}$ or $d^{\circ}_{20,w}$ of the macromolecule, from which the frictional coefficient $\langle f \rangle$ is determined. Since this is a measure of the degree of elongation of the macromolecular shape, it can be compared with the radius of gyration R_G measured by solution scattering. Sometimes the R_G is not available. This is either because trace amounts of contaminating aggregates in a sample preparation can obscure the R_G region of the scattering curve at low scattering angles, or alternatively the macromolecule is highly elongated and insufficient data at small scattering angles are available to obtain the R_G. Hydrodynamic data thus offer a most useful control of the R_G measurement by an independent method. In addition, hydrodynamic data are sometimes available for macromolecules that cannot be studied by solution scattering for reason of low solubilities or otherwise.

The author has developed the application of solution scattering to several of the complement components[4-8]. The data analyses are usually presented in terms of tri-dimensional models constructed from small spheres which give calculated scattering curves (via the Debye Equation) that are in good agreement with the experimental scattering curves, and which are compatible with electron microscopy images (Figure 1). Hydrodynamic data can be analysed in two ways. The first is to compute the degree of elongation from the frictional ratio $\langle f \rangle / \langle f_0 \rangle$, which can then be compared with the corresponding ratio of R_G / R_0 from solution scattering (Table 1), where $\langle f_0 \rangle$ and R_0 are the predicted values corresponding to the sphere of the same volume as the macromolecule under study. This is a rather limited approach. What is much more useful is the ability to compare directly the tridimensional Debye sphere model used to fit the solution scattering curve with the experimental value of $\langle f \rangle$. This will assess the degree of compatibility with hydrodynamic data. This paper will discuss how the Bloomfield- Garcia de la Torre (1977) approach[9-11] has been applied to this end. The utility of these calculations will be demonstrated, as well as the most effective way of performing them.

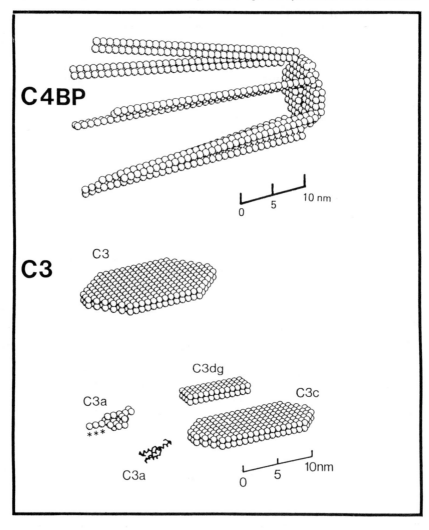

Figure 1 Debye sphere models for several complement
components and α_1 acid glycoprotein. These account for
the X-ray or neutron solution scattering curves. The
diameters of the non-overlapping spheres as shown are
listed in Table 3. Here C3 of complement and its
fragments C3a, C3c and C3dg are shown together with the
C4b binding protein. The latter interacts with the C4
family of proteins in complement, the members of which
are homologues of the C3 family.

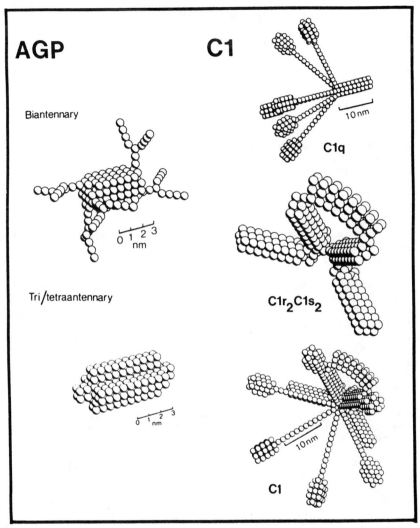

Figure 1 (cont.) α_1 Acid glycoprotein is shown. This was studied in two forms containing biantennary and tri-tetraantennary complex-type oligo-saccharides. In 1M NaCl, the former adopts an extended conformation as shown, while the latter is compact. C1q, C1r$_2$C1s$_2$ and their complex in the W-model for C1 are depicted. Note that C1r$_2$C1s$_2$ is a rod-like molecule which undergoes a major compactation when incorporated into the C1 complex and only this compact conformation is shown.

2 SURVEY OF SHAPE AND COMPOSITIONAL DATA

Four systems have been studied in detail. These provide
the basis for a discussion of the application of
Bloomfield-Garcia de la Torre calculations of frictional
coefficients. They include C3 of complement and its two
major fragments C3c and C3dg[4], C1 of complement which is
formed from two subunits C1q and $C1r_2 C1s_2$[5-7], and the C4b
binding protein (C4BP) of complement[8]. α_1 acid
glycoprotein (α_1 AGP) is an acute phase glycoprotein which
is possibly an immunosuppressive agent in plasma. α_1 AGP
has been studied in two forms with biantennary (b-) and
tri-tetraantennary (t-) carbohydrate chains in high and
low NaCl concentrations[12,13]. Tables 1 and 2 show that
the systems span a wide range of compact to highly
elongated shapes and range of molecular weights. The
carbohydrate contents range from 0% to 45%. Table 2
corresponds to the updated survey of glycoprotein volumes
presented by the author[14], and incorporates new sequence
data released since the original studies.

Qualitatively, the values of $\langle f \rangle / \langle f_0 \rangle$ and R_6 / R_0 in
Table 1 correlate with one another. A linear relation-
ship is <u>not</u> found when $\langle f \rangle / \langle f_0 \rangle$ is plotted against R_6 / R_0.
The R_6 data correspond to the mean square distance of
scattering elements from the centre of scattering mass,
while the $\langle f \rangle$ data correspond to the dynamic interactions
of the shape with its surrounding medium and are more
difficult to calculate. No quantitative correlation is
expected.

<u>Table 1</u> Summary of hydrodynamic and scattering data for
some human plasma proteins in order of increasing
molecular weight

Molecule	$s^0_{20,w}$ (S)	$\langle f \rangle / \langle f_0 \rangle$	R_6 (nm)	R_6 / R_0
b-α_1 AGP	1.8*	2.22	2.84	1.74
t-α_1 AGP	3.4*	1.35	2.24	1.31-1.37
C3dg	2.6	1.48	2.7	1.52
C3c	5.5-6.7	1.36-1.65	4.8	1.78
C3	7.3	1.54	5.1	1.71
$C1r_2 C1s_2$	8.7	2.00	17.0	4.74
C1q	10.2	2.10	12.8	3.20
C4BP	10.7-11.2	1.98-2.18	13.0	3.19
C1	15.2-16.2	1.90-2.03	12.6	2.63

* Calculated from the diffusion coefficient data in Ref.
13.

Table 2 Summary of compositional data for the plasma proteins

Molecule	Mol. Wt. ($\times 10^3$)	%CHO (by weight)	Dry volume (nm³)	Hydrated volume (nm³)	$\bar{v}$ (ml/g)
b-α_1 AGP	32	34	39	52	0.704
t-α_1 AGP	38	44	45	61	0.696
C3dg	39	0	50	66	0.743
C3c	137	2	177	233	0.739
C3	187	2	241	318	0.739
C1r$_2$ C1s$_2$	330	7	416	551	0.723
C1q	457	7	579	765	0.726
C4BP	491	12	611	811	0.717
C1	787	7	995	1316	0.725

A molecular view of the compositional data is important for the assessment of the viability of hydrodynamic calculations. The R_6 data correspond in all cases (except for C4BP, which is not significant) to the neutron radius of gyration measured at infinite contrast. This is seen as a "dry" shape. For reason of the neutron method of contrast variation using mixtures of H_2O and ^2H_2O, the hydration shell does not influence the measurement of R_6. The R_6/R_0 ratios are thus derived from the dry volumes as visualised directly from the crystal structures of proteins, amino acids and short peptides. The total volume can thus be readily computed from the compositions derived from sequence data.

In distinction, hydrodynamic data relate to hydrated glycoproteins. Some biophysicists take the view that the measured hydration depends on the technique used to measure it, and therefore cannot be rigorously defined. Most workers however assume a value of 0.3 g H_2O/ g glycoprotein. A survey of protein crystal structures in which hydration shells have been directly analysed has shown that this value is equivalent to a single monolayer of water molecules covering the macromolecular surface[14]. For this work, a good approximation in molecular terms is to visualise hydrated glycoproteins as the dry protein, surrounded by a layer of water molecules in well-defined hydrogen bonded locations, which interacts with bulk water. Bulk water is described by the most recent statistical simulations of the structure of liquid water in terms of loose fluctuating arrangements of hydrogen bonds. In this model, the hydrated volume is given by the sum of the dry volume (as above) and the volume of the

"electrostricted" bound water molecules. The latter is
0.0245 nm³, as found in a survey of hydrated small
molecule crystal structures[14]. This is less than that of
0.0299 nm³ for bulk water. These calculations lead to
good agreement between experimental and calculated
partial specific volumes v of the protein in solution.
They also account for differences in the volumes of
amino acids obtained by a variety of different physical
methods[14]. This approach has merit in being fully self-
consistent within its context.

3 A METHOD TO MODEL HYDRODYNAMIC DATA

The objective of the hydrodynamic modelling using spheres
is to approach as closely as possible the starting model
from electron microscopy, solution scattering or
otherwise with the minimum of assumptions relating to
stages in the modelling.

Hydrodynamic Theory

        The Bloomfield-Garcia de la Torre (1977) theory[9-11]
is employed to calculate frictional coefficients. This
should not be confused with the older method of
Bloomfield et al. in 1967. The Stokes Law frictional
coefficient for a sphere in a multi-sphere model is
modulated by the hydrodynamic interactions of other
spheres in the structure, and this is allowed for by use
of the Oseen tensor. Since the Oseen tensor considers
each element as a point source of friction, this is in
turn modified to allow for the finite, different sizes of
the n spheres. The frictional force $\mathbf{F}_i$ associated with
the ith element of hydrodynamic radius r_i is:

$$\mathbf{F}_i = 6\pi\eta r_i (\mathbf{u}_i - \mathbf{v}_i) \text{ for } i = 1, n \qquad (1)$$

where $\mathbf{u}_i$ is the velocity of the ith element and $\mathbf{v}_i$ is the
velocity that the solvent would have at the position of
the centre of the ith element if that element were
absent. The hydrodynamic interaction tensor $\mathbf{T}_{ij}$ when
modified for spheres of finite sizes and of different
hydrodynamic radii r_i and r_j is:

$$\mathbf{T}_{ij} = (8\pi\eta R_{ij})^{-1} \left[I + \frac{R_{ij} R_{ij}}{R_{ij}^2} + \right.$$

$$\left. \frac{(r_i^2 + r_j^2)}{R_{ij}^2} \left(\frac{1}{3} I - \frac{R_{ij} R_{ij}}{R_{ij}^2} \right) \right] \qquad (2)$$

where I is the unit vector and R_{ij} is the vector between i and j. The interaction tensor is incorporated with F_i by:

$$\mathbf{v}_i = \mathbf{v}_i{}^0 - \sum_{j=1}^{n} \mathbf{T}_{ij} \, \mathbf{F}_j \qquad (3)$$

(excluding terms with i=j), where $\mathbf{v}_i{}^0$ is the unperturbed solvent velocity when the other elements are absent.

The programs GENTRA and GENDIA implement these equations, in which iterative procedures are used to calculate $\langle f \rangle$ for an array of spheres. GENDIA utilizes the diagonal approximation, where the off-diagonal components of the Oseen tensor are usually much lower than the diagonal ones. Since GENTRA is significantly more costly in computer time, and yet gave results similar to GENDIA, GENDIA was thus used routinely for the studies described here. Its results were checked with GENTRA once the final hydrodynamic model became available. The largest change found in $\langle f \rangle$ was 1.8%

Comparison of Hydrodynamic and Scattering Models

In developing the application of these equations to the plasma proteins, two points have to be borne in mind. Very large matrices have to be diagonalised. This places computational limitations on the total number of spheres that can be used, and the resolution of the macromolecular shape is thus reduced. In addition the hydrodynamic equations are valid only for $(r_i + r_j) \leqslant R_{ij}$. Non-overlapping spheres are required in the hydrodynamic models, even though overlapping spheres are employed in the Debye models. Void spaces between the spheres are thus introduced into the model, even though these are not necessarily observed in the structure.

The Debye models for modelling the scattering curves of these proteins (Figure 1) are based on 96-636 overlapping spheres (Table 3)[4,5,7,8,13]. The spheres are situated at the centres of cubes of sides between 0.608-1.4 nm, where each sphere has the same volume as the cube. The total volume corresponds to the dry volume of the protein under study. These scattering models are sufficient to reproduce the R_G values as well as the experimental scattering curves out to structural resolutions of 4-9 nm in the C1, C3 and C4BP studies, and offer enough detail to be compared with electron microscopy images when available. To compare this with the hydrodynamic version of the model, each of its linear dimensions should ideally be larger by 2 x 0.36 nm to allow for an even distribution of the hydration shell over the surface. This takes each bound water molecule to be a sphere of diameter 0.36 nm and volume 0.0245 nm^3.

The hydrodynamic models have to allow for hydration, so the total volume of the spheres will be determined by the hydrated volume. However the void spaces between the spheres will affect the apparent dimensions of the object under study. The length (and width) of a one- (and two-) dimensional array of spheres will dominate the calculation of $\langle f \rangle$, so the values to employ can be directly estimated from the dimensions of the hydrated Debye model. The voids mean that the effective or average diameter of a line of spheres is not the same as the diameter of the equivalent solid rod. The average diameter is given by $\pi d/4$, where d is the diameter of each sphere. (By the same token, the average thickness of a two-dimensional array of spheres is also given by $\pi d/4$.) Now the volume of a line of n spheres of diameter d is given by $\pi n d^3/6$ or $0.5356 n d^3$. By comparison the volume of a solid cylinder of the same length and diameter is given by $\pi n d^3/4$ or $0.7854 n d^3$, and that of a solid cylinder of diameter $\pi d/4$ and length nd is given by $\pi^3 n d^3/64$ or $0.4845 n d^3$.

The n spheres of length nd will have the same volume as the cylinder of averaged diameter $\pi d/4$ if their diameter is increased by a factor of $\sqrt{32/3\pi^2}$ or 1.040, i.e. their volume should be increased by a factor of 1.081. If typical $\bar{v}$ values for the dry protein and the electrostricted bound water are taken as 0.75 ml/g and 0.82 ml/g respectively, the conventional hydration of 0.3 g H_2O/g protein could therefore be increased to a hypothetical figure of 0.39 g H_2O/g protein to compensate for the void spaces between the spheres in terms of a correction to the total volume. This adaptation of the total volume of the spheres is one method whereby a closer representation of the shape being modelled might be achieved, in order that it can be compared with the results of other structural investigations. Each of the triaxial dimensions of the final hydrodynamic model should be reduced by $(1-\pi/4)d$ before comparison with the Debye model is made.

One- and two-dimensional arrays of spheres are usually sufficient to model the frictional coefficient in most situations. This avoids the complications inherent in evaluating the voids inside a tri-dimensional array of spheres.

Table 3 Summary of the sphere models used to model scattering curves and hydrodynamic data

Molecule	Debye spheres (total)	d (nm)	Cal. R_6 (nm)	Exp. R_6 (nm)	Dimensions (if triaxial) (nm)		

Scattering models:

Molecule	Debye spheres (total)	d (nm)	Cal. R_6 (nm)	Exp. R_6 (nm)	Dimensions (if triaxial) (nm)		
b-α_1 AGP	175	0.608	2.81	2.84	=		
t-α_1 AGP	216	0.608	2.28	2.24	9 x	4 x	4
C3dg	96	0.8	2.96	2.7	10 x	2 x	3
C3c	348	0.8	4.99	4.8	18 x	2 x	7
C3	478	0.8	5.27	5.1	18 x	2 x	10
C1r$_2$ C1s$_2$	168	1.4	17.0	17.0	59 x	3 x	3
C1q	348	1.19	12.8	12.8	=		
C4BP	602	1.0	13.0	13.0	=		
C1	636	1.19	12.6	12.6	=		

Molecule	Hydro spheres (total)	d (nm)	Cal. $\langle f \rangle$ (x 10^{-8} g/cm/s)	Exp. $\langle f \rangle$	Dimensions (if triaxial) (nm)		

Hydrodynamic models:

Molecule	Hydro spheres (total)	d (nm)	Cal. $\langle f \rangle$ (x 10^{-8} g/cm/s)	Exp. $\langle f \rangle$	Dimensions (if triaxial) (nm)		
b-α_1 AGP	12	1.828	7.85	8.8	=		
	40	0.955					
t-α_1 AGP	22	1.78	5.96	5.6	11 x	5 x	5
C3dg	5	3.004	6.39	6.4	15 x	3 x	3
C3c	18	3.004	10.2	8.9-10.9	21 x	3 x	9
C3	24	3.004	11.3	11.2	21 x	3 x	12
C1r$_2$ C1s$_2$	13.5	4.376	17.8	17.5	59 x	4 x	4
C1q	66	2.004	20.3-21.0	20.5	=		
	6	5.474					
C4BP	91	2.55	21.5	19.7-21.6	=		
	1	5.46					
C1	66	2.004	21.5	22.3-23.7	=		
	6	5.474					
	12	4.55					

4 COMPARISONS WITH DEBYE SCATTERING MODELS

The method of calculating hydrodynamic models in this paper is an improved procedure compared to that initially developed for α_1 AGP, C1, C3 and C4BP. The original simulations were thus repeated and compared. The improvements are enumerated: (a) The survey of dry and hydrated volumes by the author[14] has clarified how the total molecular volume and $\bar{v}$ are best calculated from sequence data; (b) More direct comparisons can be made

with the dimensions of the original starting model from
the Debye simulations on the basis of the above
discussion of the effect of hydration and the void spaces
in the hydrodynamic model; (c) Upgrades on the CDC Cyber
855 used for this work meant that if required up to 150
spheres can now be employed in place of the earlier limit
of 112 spheres (although here this did not turn out to be
necessary). The use of as many spheres as possible means
that the shape can be more closely approximated to that
derived by other physical methods. The final models are
summarised in Figure 2.

Hydrodynamic modelling of C1q and the C1 complex

The C1q subcomponent is structurally the best
understood of the systems shown in Tables 1-3. Sequence
studies and electron microscopy have shown that it is
composed of six globular heads that are connected by six
collagenous arms or stalks to a central base (Figure 1).
Calculations based on the known pitch of the collagen
triple helix and on solution scattering data show clearly
that the length of the collagenous arm between the head
and the base is close to 14.5 nm, and not to 11.5 nm as
initially proposed from electron microscopy studies[6,7].
The solution scattering data also showed that the average
arm-axis angle in C1q is $40°-45°$. Unfortunately the
initial calculations of $s^0_{20,w}$ with this angle gave a
value of 9.03 S, which was 11% less than the accepted
experimental $s^0_{20,w}$ value of 10.2 S. The calculations
were not able to choose clearly between the two possible
arm lengths. Recalculation according to the above
procedure resolved this discrepancy, giving a final
calculated $s^0_{20,w}$ of 10.0-10.3 S for an arm-axis angle of
$40°-45°$. This improvement was primarily due to the use of
a $\bar{v}$ value based on the hydrated volume of C1q, and not
the dry one, in the calculation of $\langle f \rangle$. The importance of
this is readily seen from the expressions for the
frictional coefficient $\langle f \rangle$:

$$\langle f \rangle = \frac{M_r \ (1 - \rho \ \bar{v})}{N_A \cdot s^0_{20,w}} = \frac{kT}{d^0_{20,w}} = 6\pi\eta a \qquad (4)$$

where M_r is the molecular weight of the hydrated solute,
ρ is the solvent density, N_A is Avogadro' Number, k is
Boltzmann's constant, $s^0_{20,w}$ and $d^0_{20,w}$ are the
sedimentation and diffusion coefficients, T is the
absolute temperature, η is the solvent viscosity, and a
is the Stokes radius. Note that the calculation of $\langle f \rangle$
from $s^0_{20,w}$ data requires accurate $\bar{v}$ values since the
difference within the bracketed term leaves a small

remainder which is sensitive to $\bar{v}$.

$Clr_2 Cls_2$ has been studied by electron microscopy and by solution scattering to show that it has a rod-like extended shape in solution of length 59 nm[5]. Hydrodynamic simulations, initially using several expressions for cylindrical models[5,7], and more recently using spheres[7], were able to reproduce the $s^o_{20,w}$ value of 8.7 S by a cylinder of length 59.0 nm or spheres of length 69.4 nm. Since then, sequences have become available for the Cls and Clr subunits of $Clr_2 Cls_2$ to show that the molecular weight used in these earlier studies was too high by 14-18%. The recalculation shows that a line of 13.5 spheres of length 59 nm yields a $s^o_{20,w}$ of 8.6 S, which is in very good agreement with the value of 8.7 S by experiment, with a length that is highly compatible with that determined from electron microscopy and neutron scattering.

The simulations were extended to the calculation of possible models for the association of $Clr_2 Cls_2$ with Clq to form C1[7]. Table 1 shows from both the $\langle f \rangle / \langle f_0 \rangle$ and the R_G / R_0 ratios that C1 is more compact than Clq. It can be shown from considerations of the R_G and the Clq structure that a shape change in Clq is not able to account for this compactation, and occurs instead in $Clr_2 Cls_2$. Several models have been proposed for the shape of bound $Clr_2 Cls_2$ within C1, however only the W-model derived from Debye simulations is discussed here. By placing one-quarter segments of the $Clr_2 Cls_2$ model in a compact arrangement onto four adjacent arms of the Clq structure (Figure 1) which has a 40° arm-axis angle, the simulation was able to account for the scattering curve out to a nominal structural resolution of 8 nm. The corresponding hydrodynamic model lead to a simulated $s^o_{20,w}$ of 16.8 S. Given the complexity of the C1 structure, this·is in fair agreement with the experimental values of 15.2-16.2 S. The difference between the calculated and experimental values may reflect the possibility that the C1 complex is slightly more elongated in its structure than that depicted.

Hydrodynamic modelling of C3 and its fragments

C3 of complement is initially converted by the removal of a short peptide C3a to yield the active form C3b. This is eventually degraded by other complement components to the major fragments C3c and C3dg. The solution scattering method was able to approximate the shapes of C3, C3c and C3dg by lamellar ellipsoids, and to

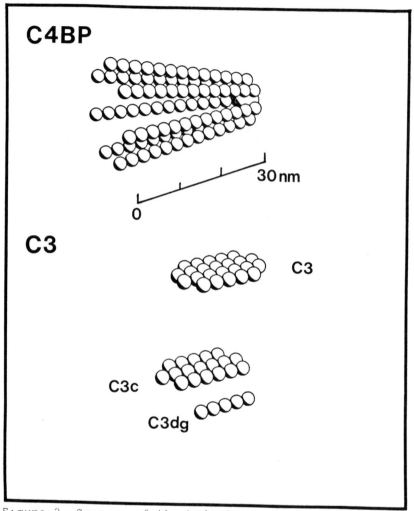

<u>Figure 2</u> Summary of the hydrodynamic models that were developed. These offer improved agreements with the experimental <f> values than those developed earlier. Details of the models are summarised in Table 3, and they follow the order as presented in Figure 1. Hydrodynamic models are shown for the C4b binding protein, with an arm-axis angle of 10°, and for C3, C3c and C3dg, showing how C3c and C3dg are constituted as distinct domains in C3. C3a, which is 5% of the mass of C3, is not directly considered in these models. All four models are drawn to the same scale.

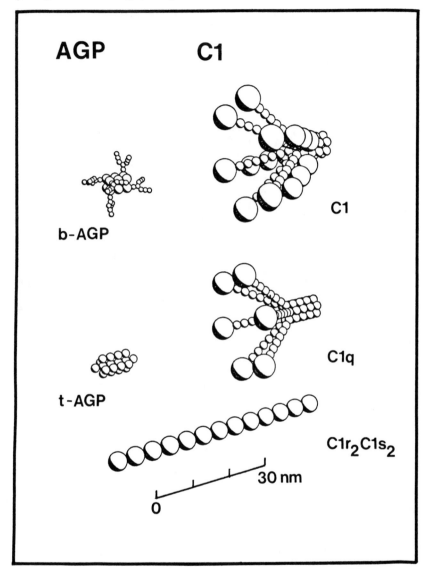

<u>Figure 2</u> (cont.) Hydrodynamic models for α_1 acid glycoprotein in its two forms with biantennary and with tri-tetraantennary carbohydrate chains, and for C1q and $C1r_2 C1s_2$ and their complex to form C1. The arm-axis angle of C1q is 40°; $C1r_2 C1s_2$ is shown in an extended shape. All five models are drawn to the same scale.

show that it was possible to build the structure of C3 from the side-by-side positioning of C3c and C3dg[4]. The hydrodynamic simulations were used to show whether or not the $s^0_{20,w}$ data in the literature were compatible with such a model of domain assembly in the structure of C3.

As for C1q and C1r$_2$C1s$_2$, more satisfactory calculations were performed here that are based on more precise sequence and volumetric data. For C3, the Debye dimensions of 18.4 x 1.6 x 10.4 nm are equivalent to hydrated dimensions of 19.1 x 2.3 x 11.1 nm. The final hydrodynamic model of 24 spheres had dimensions of 21 x 3 x 12 nm, which is in very satisfactory agreement with the scattering model. Note that its average thickness is 2.4 nm. Table 3 shows that the simulated and experimental ⟨f⟩ values are in very good accord. This model was subdivided into those for the two fragments with 18 and 5 spheres. These lead to calculated ⟨f⟩ values that are also in very good agreement with experiment. The hydrodynamic simulations are therefore fully consistent with the domain model proposed from solution scattering, although comparisons of the dimensions in Table 3 show that the hydrodynamic model for C3dg is somewhat longer than the Debye model.

Hydrodynamic modelling of C4BP

The C4b binding protein (C4BP) provides a case study in which its R_G had been obtained at the angular limit of the scattering curve as seen in synchrotron X-ray solution scattering[8]. It was necessary to confirm this R_G value by a comparison with experimental $s^0_{20,w}$ data.

The C4BP has a multisubunit structure composed of seven arms attached at a base (Figure 1). The evidence for this came from electron microscopy, which however suggested that the seven arms were splayed out in all directions. The construction of Debye models with arms of length 33 nm as seen in the micrographs showed that such a model with an arm-axis angle of 90° leads to a predicted R_G of 21 nm. This is markedly different from the observed R_G of 13 nm. Agreement with this R_G could however be found with an arm-axis angle of 5°-10° instead of 90°. Hydrodynamic simulations of the $s^0_{20,w}$ were performed as a function of the arm-axis angle. Figure 3 showed that these fully verified the premise that the arm-axis angle in C4BP in solution is of the order 5°-10°. Indeed, these models went on to account successfully for the X-ray scattering curve to a resolution of 6.4 nm.

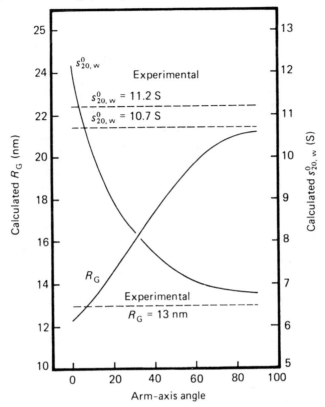

<u>Figure 3</u> Dependence of the R_G and $s^0{}_{20,w}$ values for C4BP on the arm-axis angle of C4BP. From Ref. 8.

The C4BP hydrodynamic model was redeveloped here by assigning the volume of the base as the hydrated volume of seven times the C-terminal 58 residues (which is free of carbohydrate). The hydrated volume remaining was divided between the seven arms of length 33 nm each. Despite the differences in the sphere volumes, very similar $\langle f \rangle$ values to the earlier work were obtained. This is because the arm length was kept constant. The change of $\bar{v}$ from 0.740 ml/g previously to 0.717 ml/g here however causes the $s^0{}_{20,w}$ to change from 10.1 S to 10.8 S. The latter figure is in better agreement with the observed values of 10.7 S to 11.2 S.

Hydrodynamic modelling of α_1 acid glycoprotein

The interest in α_1 acid glycoprotein (α_1AGP) is that this is the most heavily glycosylated plasma protein that is known. In addition to this, neutron scattering experiments discovered an unusual conformational change in native α_1AGP[1,2]. This underwent an expansion on passing from 0.2M NaCl buffers to 1.0M NaCl buffers. Neutron contrast variation was used to demonstrate that the hydrophobic part of the structure remained unchanged, while the hydrophilic part (including the carbohydrate moities) had changed. Native α_1AGP could be separated into two fractions, one containing biantennary oligosaccharide groups, and the other containing tri- and tetraantennary groups. The conformational change could only be detected for the biantennary form on passing to high salt, not for the tri-tetraantennary form[13]. This result was confirmed by laser light scattering experiments which confirmed that the two forms had different diffusion coefficients $d^0_{20,w}$ of 4.59 x 10^{-7} cm^2/s and 7.19 x 10^{-7} cm^2/s in 1M NaCl buffers.

Debye and hydrodynamic models were developed to show whether or not the R_G and $\langle f \rangle$ parameters could be explained in terms of a compact core to which carbohydrate chains were attached. The hydrodynamic calculations were improved by allowing for hydration, and by selecting a shape for the protein core that is closer to the Debye dimensions after increasing each one by 0.72 nm.The agreements shown in Table 3 showed that this molecular model was reasonable. Given the very high carbohydrate content of α_1AGP, it was gratifying that the methodology was satisfactory. In general, glycoproteins have been thought to be associated with much higher hydration levels than 0.3 g H$_2$O/ g glycoprotein. The feasibility of both types of modelling calculations, based on views of the "dry" and "hydrated" glycoprotein, suggests instead that explicit models based on a conventional hydration of 0.3 g H$_2$O/ g glycoprotein and showing the carbohydrate chains to extend deeply into the surrounding solvent are able to offer quantitative insights into this aspect of glycoprotein structure.

5 CONCLUSIONS

The original hydrodynamic calculations on which this review is based[4,7,8,13] were the first successful application of the Bloomfield-Garcia de la Torre method to plasma glycoproteins, whose molecular weights range from 38,000 to 832,000. Reasonable agreements could be readily obtained. The refinements proposed here for the revised calculations, which take into account the survey of volumes and hydration shells, lead to improved agreements between the experimental and simulated frictional coefficients in 7 of the 9 cases. The average discrepancy with experiment has been more than halved. The poorest agreements were obtained with the least well-defined shapes, where $b-\alpha_1$ AGP and the C1 complex may be more elongated than anticipated in the simulations. Excluding these two, it is reasonable to expect that the calculated and experimental $\langle f \rangle$ agree to within ± 0.3 x 10^{-8} g/cm/s. In the five triaxial models of Table 3, comparisons of the Debye and hydrodynamic dimensions after correction to the former for hydration and the latter for voids (see text) leads to a r.m.s. difference of 1 nm between the two models. Most notably, the calculation of $\langle f \rangle$ for C1q of complement is in complete agreement with the experimental data unlike previously. This example emphasizes the importance of using the correct $\bar{v}$ values to extract $\langle f \rangle$ from $s^0_{20,w}$.

It has been shown how a generally accepted view on protein hydration can be successfully incorporated in the simulations. Use of the hydrated volume in place of the dry volume for the total volume of the spheres leads to a reduction in the number of spheres. The use of hydration levels above 0.3 g H_2O/g protein implies that the spheres should be replaced by more complex assemblies of smaller spheres. This minimizes difficulties in terms of the voids between the spheres and matching their overall dimensions to the shape being modelled. Finally it should be noted that different proteins may have different levels of hydration[14]. This introduces a further unknown into the simulations, however the agreements obtained in the present study suggest that its influence is not significant.

In all these simulations, it is important to remember that the calculations do not lead to unique solutions. The $\langle f \rangle$ data give only one shape parameter, and many other models with the same total volume will give similar $\langle f \rangle$ values. The shape calculations have to be constrained by the use of as many independent lines of structural evidence as possible. In ideal circumstances, this leads to one structural unknown, which can then be determined by the hydrodynamic method. The C4BP study exemplifies this, where the $s^0_{20,w}$ data were used to support an R_G value whose determination was not totally unambiguous.

One limitation of the calculations is that the hydrodynamic models are static ones, and that macromolecular flexibility has not been explicitly considered. Most of the plasma proteins under consideration here are however known or are expected to possess flexible structures in which distinct structural domains are free to move relative to one another. In principle, some insight on this is available in the cases of C1q and C4BP in that the calculation of $\langle f \rangle$ as a function of the arm-axis angle of the molecule shows the range in which $\langle f \rangle$ can vary. The experimental $\langle f \rangle$ could then be compared with a $\langle f \rangle$ that has been calculated from the application of a Boltzmann or Gaussian weighting scheme to a family of related conformations of C1q or C4BP.

Finally the immunological relevance of the hydrodynamic properties of the components of the complement cascade and of IgG and IgM antibodies should be remembered. The role of these macromolecules is to recognize and destroy foreign particles. Much evidence from sequences, crystallography, electron microscopy, hydrodynamic data and solution scattering has shown that many of these macromolecules are constructed of repeated structural motifs that are arranged to yield highly elongated structures. Thus IgG is formed from the assembly of 12 Ig folds into a three-armed structure, IgM has 70 Ig folds arranged as a discoid pentamer with 10 distinct arms, C4BP has 56 short consensus repeats contained within a seven-armed structure, etc. The effect of these very elongated structural forms is to slow down the rate of diffusion of these macromolecules while they are being transported in the bloodstream. Control of the rate at which complement is activated is essential to maintain the selectivity of the immune response to an invading foreign substance, and to prevent the unrestrained consumption of the available complement components in the blood. This taken with the elongated structures of the complement components implies that the reduced diffusability of these elongated macromolecules offers a physical mechanism for the control of complement activation by restricting it to the volume immediately adjacent to the foreign substance.

ACKNOWLEDGEMENTS

S.J.P. gratefully acknowledges support from the Wellcome Trust. The manuscript was completed on 3.8.1988.

REFERENCES

1. K.B.M. Reid, Essays in Biochemistry, 1986, 22, 27.
2. S.J. Perkins, Biochem. J., 1988, 254, 1.
3. S.J. Perkins, 'New Comprehensive Biochemistry' (eds. A. Neuberger & L.L.M. Van Deenan), Elsevier, Amsterdam, 1988. Vol. 18B. Chapter 6, p. 143.
4. S.J. Perkins and R.B. Sim, Eur. J. Biochem., 1986, 157, 155.
5. J. Boyd, D.R. Burton, S.J. Perkins, C.L. Villiers, R.A. Dwek and G.J. Arlaud, Proc. Natl. Acad. Sci. U.S.A., 1983, 80, 3769.
6. S.J. Perkins, C.L. Villiers, G.J. Arlaud, J. Boyd, D.R. Burton, M.G. Colomb and R.A. Dwek, J. Mol. Biol., 1984, 179, 547.
7. S.J. Perkins, Biochem. J., 1985, 228, 13.
8. S.J. Perkins, L.P. Chung and K.B.M. Reid, Biochem. J., 1986, 233, 799.
9. J. Garcia de la Torre and V.A. Bloomfield, Biopolymers, 1977, 16, 1747.
10. J. Garcia de la Torre and V.A. Bloomfield, Biopolymers, 1977, 16, 1779.
11. J. Garcia de la Torre and V.A. Bloomfield, Quart. Rev. Biophys., 1981, 14, 81.
12. Z.Q. Li, S.J. Perkins and M.H. Loucheux-Lefebvre, Eur. J. Biochem., 1983, 130, 275.
13. S.J. Perkins, J.P. Kerckaert and M.H. Loucheux-Lefebvre, Eur. J. Biochem., 1985, 147, 525.
14. S.J. Perkins, Eur. J. Biochem., 1986, 57, 169.

16

Flow Properties of Proteoglycan Solutions

By T. E. Hardingham,[1] C. Hughes,[1] V. C. Mow,[2] and W. M. Lai[2]

[1]KENNEDY INSTITUTE OF RHEUMATOLOGY, HAMMERSMITH, LONDON, U.K.

[2]ORTHOPAEDIC RESEARCH LABORATORY, COLLEGE OF PHYSICIANS AND SURGEONS, COLUMBIA UNIVERSITY, NEW YORK, U.S.A.

Introduction

Proteoglycans are complex macromolecules with an expanded branched structure composed of a protein backbone to which are attached over 100 long carbohydrate side chains (chondroitin sulphate and keratan sulphate) (Fig 1)[1]. These are polyanionic polysaccharides containing regular carboxylate and sulphate groups. Each proteoglycan contains a specific site for aggregation with hyaluronate. This involves a globular protein domain at one end of the proteoglycan which interacts with a short segment (10 monosaccharides) of the hyaluronate chain ($Kd\sim10^{-8}M$) (Fig 1a). Hyaluronate is unbranched and of great length (1-5μ) such that up to 100 or more proteoglycans may bind to each chain. This forms large aggregates in which each proteoglycan-hyaluronate bond is further stabilised by a globular link protein. The link protein greatly increases the strength of the PG-HA bond and dissociation is no longer detected[2]. This mode of aggregation creates large molecular assemblies, but they are of finite size and do not show cross-linking or gel forming properties (Fig 1b).

Proteoglycans of this type are abundant in vertebrate cartilage and are responsible for the elastic and compressive properties of this tissue. All the components of the proteoglycan aggregate are synthesised by the chondrocytes within the tissue and aggregation is an extracellular process that immobilises the proteoglycans within the extracellular matrix[3]. The structure of cartilage contains a dense network of fine collagen fibres that provide shape and

1a

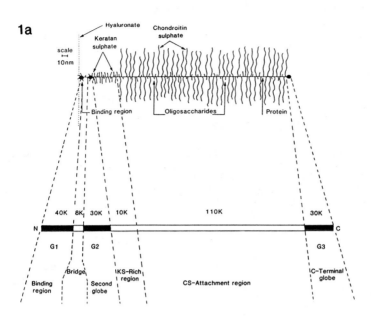

1b

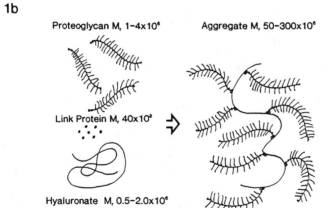

Fig 1 a) Schematic Model of Proteoglycan Structure and Protein Core Domains.

b) Proteoglycan Aggregation with Hyaluronate and Link Protein.

form to the tissue. The proteoglycans pack the space
between the fibrillar network and, because of the
polyanionic charge they create a large osmotic pressure
that draws water into the tissue. This hydrates the
interfibrillar space, expands the collagen network and
places it under tension. The tissue is thus stiffened
and made resiliant inspite of a high water content
(70-80% w/w). Proteoglycans are present in the tissue
at 50-80mg/ml and are maintained at this high
concentration by biosynthesis in the chondrocytes.

Preparation and Properties of Proteoglycan Solutions

The proteoglycans are extracted from cartilage in
4M guanidine HCl. This dissociates the aggregates,
which reform on lowering the guanidine HCl
concentration. Preparations containing reformed
proteoglycan aggregates or purified proteoglycan
monomers are obtained by CsCl density gradient
centrifugation. The ability to isolate purified
monomeric proteoglycan separate from link protein and
hyaluronate enables aggregates of different composition
to be formed by remixing the components in varying
proportions in 4M guanidine HCl, followed by re-
equilibration to associative conditions[2].

The proteoglycans are highly soluble in aqueous
solutions and in 0.15M NaCl much of the electrostatic
interactions of the polyanionic chains are abolished.
In dilute solution below 1mg/ml their molecular
characteristics can be determined. Analytical
ultracentrifugation shows the sedimentation coefficient
of monomers ($S^0_{20}w$ 24) and aggregates ($S^0_{20}w$ 80-120)[2].
These properties show a strong negative concentration
dependance which results from intermolecular
interactions. It is particularly marked for aggregates
such that at above 6mg/ml the sedimentation coefficient
is not distinguishable from that of the monomer at the
same concentration. At concentrations approaching those
found in cartilage (50-80mg/ml) the solution properties
are dominated by the network of intermolecular
interactions.

The compressive loading of cartilage involves
matrix deformation and fluid flow between the fibres.
In this project the flow properties of proteoglycan
solutions at concentrations approaching those found in
cartilage were investigated in order to understand more
of how they contribute to the dynamic properties of the
tissue.

Figure 2

CONE-ON-PLATE VISCOMETER

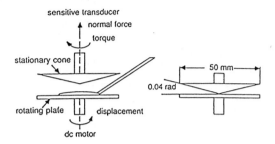

Figure 3

Dynamic Shear Test

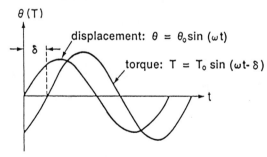

Some evidence has shown that proteoglycans have less expanded structures at high concentration than when in dilute solution[4]. This appears to be caused by intermolecular interactions between side chains which lead to some contraction of the branched structure, such that it is less extended and hydrodynamically smaller. This was shown by gel permeation chromatography where the apparent hydrodynamic size of a radiolabelled sample was smaller when determined in the presence of a high concentration of a high molecular weight dextran in the supporting solvent. The limiting viscosity number was also reduced by up to 50% in a concentration dependant way by having other long chain polymers, dextran or chondroitin sulphate chains, present in the solution[4]. Inspite of this contraction of the proteoglycan domain at high concentrations, calculations from the limiting viscosity number would suggest that proteoglycans in cartilage have available to them only a fraction of their normal volume and there must be considerable overlap of the molecular domains.

Rheological Properties

In order to investigate the properties arising from the network of intermolecular interactions, solution of proteoglycan monomers and aggregates were tested at various concentrations in a cone or plate viscometer (Rheometrics RMS 800) (Fig 2). With this equipment the complex dynamic shear modulus was determined from measurements on small amplitude (0.02rad) sinusoidal shear oscillations at different frequencies(1 to 20 Hz), and the apparent viscosity and normal stress difference (the force parallel with the axis of rotation) were determined from measurements at constant shear rates (0.25 to 250 s^{-1})[5][6].

The results were analysed using an Oldroyd four-parameter non-linear rheological equation of state. This was chosen because of its simplicity and its ability to adequately describe the major features of the properties of these solutions[5].

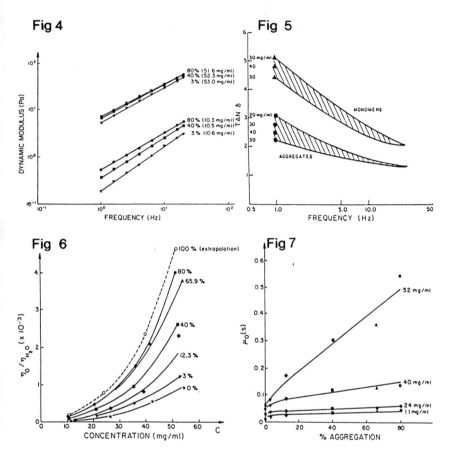

Fig 4 Dynamic Modulus of Proteoglycan Solutions at ~10mg/ml and ~50mg/ml and Containing Different Proportions of Aggregate (3%-80%).

Fig 5 Variation of Tan δ with Frequency for Solutions of Proteoglycan Monomers and Aggregates.

Fig 6 Variation of Zero Shear-Rate Viscosity Coefficient with Proteoglycan Concentration and with Different Proportions of Aggregates.

Fig 7 Variation of Non-Linear Viscosity Parameter with the Proportion of Aggregate in Proteoglycan Solutions.

For Steady-State Oscillatory Shear

Dynamic shear modulus $G^* = n_0 \omega \left[\dfrac{1+(\lambda_2\omega)^2}{1+(\lambda_1\omega)^2} \right]^{1/2}$

$\qquad$ ω $\qquad$ frequency oscillation
$\qquad$ n_0 $\qquad$ zero shear rate viscosity
$\qquad$ μ_0 $\qquad$ non-linear viscosity parameter
$\qquad$ λ_1 $\qquad$ relaxation time
$\qquad$ λ_2 $\qquad$ retardation time

For Constant Shear Rate (Steady Flow)

Apparent viscosity $n(\dot{\gamma}) = n_0 \; \dfrac{1+\mu_0\lambda_2\dot{\gamma}^2}{1+\mu_0\lambda_1\dot{\gamma}^2}$

Normal stress $\qquad$ $n_1(\dot{\gamma}) = \qquad \dfrac{2n_0\dot{\gamma}^2(\lambda_1-\lambda_2)}{1+\mu_0\lambda_1\dot{\gamma}^2}$
function

$\qquad$ $\dot{\gamma}$ $\qquad$ shear rate

Infinite shear rate viscosity $n_\infty = n_0 \; \dfrac{(\lambda_2)}{(\lambda_1)}$

Comparisons were made between solutions of pure monomers and solutions containing varying proportions of aggregate[5, 6]. These were tested in a concentration range 10-50mg/ml. The results showed that aggregates have a much higher dynamic modulus, they appear to store more energy because of stronger network formation (Fig 4). This is evident from the determination of the phase angle of the solution (Fig 5):

$$\tan\delta = \frac{G_2 \text{ (loss modulus)}}{G_1 \text{ (storage modulus)}}$$

It was found that $\tan\delta$ was lower in aggregated proteoglycan solutions and at higher concentrations (Fig 5). It also decreased for monomer and aggregate as frequency increased. Aggregated proteoglycan at high concentration thus showed a higher ratio of stored energy to that dissociated during harmonic shearing of the solution.

The zero-shear rate viscosity of aggregates is also much larger than that of monomers showing a more ordered structure at rest (Fig 6), but pronounced

Fig 8

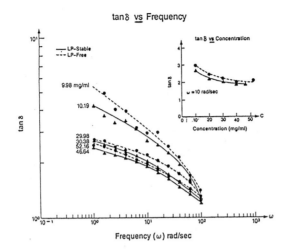

Fig 9

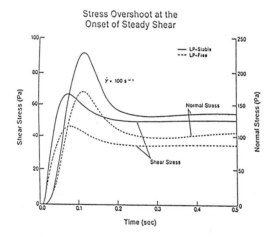

Fig 8 Comparison of Link-Stable and Link-Free Preparations of Proteoglycan Aggregates. Variation of Tan δ with Frequency and Effect of Concentration (inset).

Fig 9 Comparison of Link-Stable and Link-Free Preparations of Proteoglycan Aggregates. Stress Overshoot at the Onset of Steady Shear.

shear-thinning showed some disruption of this even at low shear rates. The properties of aggregates were more highly concentration dependent than those of monomers and therefore showed a larger non-linear viscosity parameter (Fig 7).

Effect of Link-stabilization on Rheological Properties

Having shown major difference in properties between solutions containing aggregates compared with those with only monomers, it was important to show the contribution made by link protein stabilisation to these properties. Two preparations were thus compared in which only one was link stabilised[7]. Over the concentration range investigated (10-40mg/ml) which corresponded to a molarity of proteoglycan of $6 \times 10^{-6}M$, it was calculated that at equilibrium the binding of proteoglycan to hyaluronate $(K \sim 2 \times 10^{-8}M)$ would be about 99% associated. There should therefore be a similar proportion of aggregates in each preparation. Comparison of the properties showed link-stable aggregate to show stronger network formation with a higher viscosity and a lower tan δ reflected more energy stored compared with that lost during sinusoidal shear (Fig 8). Investigation of time dependant changes such as the stress growth at the onset of steady shear, which provides a measure of the energy required in disrupting the existing network, showed link stable structures to be stronger (Fig 9), as did stress relaxation determinations at the cessation of steady shear.

Conclusions

Proteoglycan aggregation provides an extracellular mechanism for the assembly of proteoglycans into supramolecular structures in connective tissue matrices. The function in cartilage appears to be to immobilise the proteoglycans at high concentration within the collagen network. Investigation of the rheological properties showed that the concentration of proteoglycans, their formation into aggregates, the size of the aggregates and their stabilisation by link protein all contributed separately and additively to the network properties present in solution. It is notable that the dynamic modulus of cartilage in shear is $\sim 1 \times 10^6 Pa$ [8,9] whereas that of concentrated proteoglycan is only $\sim 1.0Pa$. The function of proteoglycan is not therefore to provide shear stiffness, but to organise and immobilise the collagen framework in an extended state so that the collagen fibrils are tensed during shear deformation. The viscoelastic properties contributed by proteoglycans within the cartilage matrix were found to be particularly dependant on their concentration and on their organisation into aggregate. These two factors are closely interrelated as aggregates immobilise proteoglycan within the tissue and thereby helps to maintain the high concentration. Factors influencing the loss of aggregation or otherwise promoting a fall in proteoglycan concentration in the tissue will thus have damaging effects on the load-bearing properties of articular cartilage.

Acknowledgements

This work was support by the Arthritis and Rheumatism Council (UK) and NIH grant AR 38742 (USA). Figs 4, 6 and 7 are taken from Hardingham et al. (1987), Fig 5 from Mow et al. (1984) and Figs 8 and 9 from Zhu et al. (1987) with the permission of the authors and copyright holders.

References

1. T.E. Hardingham, Biochem.Soc.Trans., 1981, 9, 489.
2. T.E. Hardingham, Biochem.J., 1979, 177, 237.
3. T.E. Hardingham, 'The Control of Tissue Damage', Elsevier, Amsterdam, 1988, Vol. 15, p.41.
4. G.S. Harper and B.N. Preston, J.Biol.Chem., 1987, 262, 8088.
5. V.C. Mow, A.F. Mak, W.M. Lai, L.C. Rosenberg and L.H. Tang, J.Biomech., 1984, 17, 325.
6. T.E. Hardingham, H. Muir, M.K. Kwan, W.M. Lai and V.C. Mow, J.Orthop.Res., 1987, 5, 36.
7. W. Zhu, W.M. Lai, V.C. Mow, C. Hughes, T.E. Hardingham and H. Muir, Proc. 2nd Japan/USA/China Biomechanics Conference, 1987.
8. W.C. Hayes and A.J. Bodine, J.Biomech., 1978, 11, 407.
9. V.C. Mow, M.H. Holmes and W.M. Lai, 'Biomechanics: Principles and Applications', Martinus Nijhoff, The Hague, 1982, p.47.

17

Models for the Macromolecular Structure of Mucus Glycoproteins

By J. K. Sheehan[1] and I. Carlstedt[2]

[1]DEPT. OF BIOCHEMISTRY AND MOLECULAR BIOLOGY, THE MEDICAL SCHOOL, UNIVERSITY OF MANCHESTER, MANCHESTER M13 9PT, U.K.
[2]DEPT. OF PHYSIOLOGICAL CHEMISTRY 2, P.O. BOX 94, S-221 00 LUND, SWEDEN.

1 INTRODUCTION

The characteristic viscous and sticky properties of mucus, whether from slug, earthworm, fish or nose, need no introduction even to the layman. Regardless of the source, the general functions of this substance are in many respects very similar and usually involve the protection of a delicate mucosal surface from the physical, chemical and biological assaults of the outside world. Saliva lubricates food, allowing it to be chewed and swallowed, gastric mucus shields the stomach lining from physical and chemical erosion, bronchial mucus facilitates the removal of detritus from the lung whereas cervical mucus guards the entrance to the womb and plays a vital role in fertility owing to the variation of its properties during the ovulatory cycle. However, the detailed mechanisms by which these and other functions are implemented remain obscure.

Mucus is a complex mixture but the components primarily responsible for its visco-elastic properties are the mucus glycoproteins or the 'mucins'. It is obvious that a detailed understanding of the biology of mucus must be preceded by a clear perception of the structure and properties of these macromolecules. However, there is some disagreement as to their 'architecture' and here we will discuss the proposed structural models in the light of our own studies on cervical, bronchial and gastric mucins.

Proposed Models for Mucus Glycoproteins

Mucus is not easily solubilised and this has lead to diverse approaches for extracting mucins such as slow stirring in denaturing solvents, high-shear homogenisation, reduction of disulphide bonds and digestion with proteinases[1]. Consequently, the properties of the purified product have varied substantially but certain structural features (Figure 1) are agreed upon, one being the presence of large proteinase–resistant glycosylated regions (oligosaccharide 'clusters') of the protein core. The oligo-saccharides usually contain 2–10 monosaccharides; typically fucose, galactose, $\underline{N}$–acetylglucosamine, $\underline{N}$–acetylgalactosamine and sialic acids. The glycans may be branched or linear and are linked to threonine or serine in the protein core via $\underline{O}$–glycosidic linkages. The presence of 'naked' protein regions susceptible to proteinase digestion is generally recognized, as is the importance of disulphide bonds in forming larger mucin structures. However, the precise arrangement of the oligosaccharide 'clusters' in the macromolecule differs between the proposed structural models.

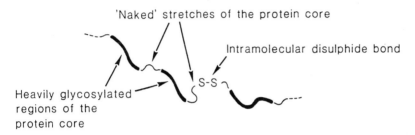

'Naked' stretches of the protein core

Intramolecular disulphide bond

Heavily glycosylated regions of the protein core

Figure 1   Schematic drawing showing some of the agreed structural features of mucins

'Windmill' Model. In a series of papers, Allen and collabo-rators described the purification and characterisation of pig gastric mucins and a number of sub–fragments obtained after reduction of disulphide bonds as well as digestion with pro-teinases[2-10]. The model emerging was called the 'windmill' model and is schematically represented in Figure 2a. The molecule (M_r approx. 2–3x10[6]) is envisaged as having four heavily glyco-sylated subunits (M_r approx. 500 000) joined by disulphide bonds to a central 'linking' protein with M_r 70 000.

'Swollen coil array' Model. This model envisages mucins as essentially linear macromolecues which are coiled into an overall spheroidal conformation. The 'basic units' themselves are imagined as spheroidal swollen oligosaccharide clusters joined by extended less glycosylated regions of the protein core (Fig. 2b). The model seeks to explain the large amount of proline in the oligo-saccharide 'clusters', the hydrated and spherical configuration of these structures in solution and some observations in the electron microscope[11-13].

'Lectin' Model: This model, which was recently proposed by Silberberg[14], is depicted in Figure 2c. It focuses on the observation that a heavily glycosylated 'basic unit' with M_r approximately 500 000 is obtained by many investigators after treating mucins from various sources with disulphide bond cleaving agents and/or proteinases. In brief, it postulates a mechanism by which 'basic units' elongate into larger structures by strong but non-covalent interactions between a lectin-like protein 'site' at one end and a specific sugar sequence at the opposite end.

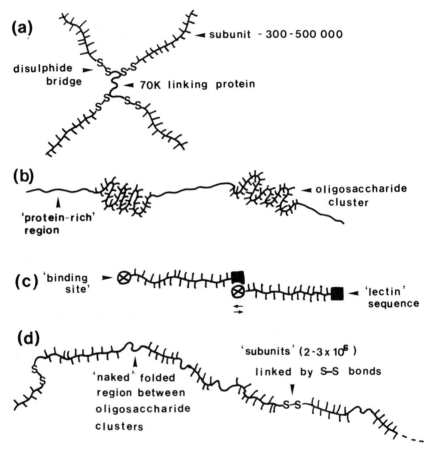

Figure 2   Schematic illustration of different proposed models for the macromolecular structure of mucins. a) 'windmill' model, b) 'swollen coil array' model, c) 'lectin' model, d) 'linear random coil' model.

'Linear Random Coil' Model. This model envisages the molecules as long filamentous particles (Figure 2d) made up from 2–10
subunits (M_r approx. 2×10^6), each containing 4–5 heavily glycosylated domains (M_r 300–500 000) joined by stretches of 'naked'
protein. The latter structures are folded and stabilised by
disulphide bonds. The model, proposed by Carlstedt and Sheehan,
is based on hydrodynamic[15-18] and electron microscopy data[19] and
differs from that of other workers mainly by the larger size of
the intact mucins and, in particular, the subunits. In this paper
we review our evidence for this model, comparing mucins purified
from the gel phase of chronic bronchitic (CBM) sputum, cervical
pregnancy (CPM) and pig gastric mucus (PGM).

2 ISOLATION AND PURIFICATION

Our philosophy for the isolation and purification of mucins has
followed the work on cartilage proteoglycans of Sajdera and
Hascall[20]. The principle of their approach was to solubilise the
macromolecules by disruption of non–covalent interactions while
protecting them against degradation caused by high shear and
endogenous proteolytic activity. The mucus samples, whether
gastric, bronchial or cervical, are collected rapidly and stored
frozen. Mucus is extracted by slow stirring in 6M guanidinium
chloride containing proteinase inhibitors such as di–isopropyl
phosphorofluoridate (DFP) and EDTA for the inhibition of serine–
and metalloproteinases, respectively, and $\underline{N}$-ethylmaleimide (NEM)
for the inhibition of thiol proteinases as well as for decreasing
the potential for thiol exchange reactions. This procedure
solubilises 80–90% of the glycoproteins from cervical and bronchial mucus and 60–70% from gastric mucus. Non–soluble components
are removed by centrifugation at this stage and purification is
achieved by isopycnic density–gradient centrifugation in CsCl/4M
guanidinium chloride with an initial density of approximately
1.40 g/ml (Figure 3a). The mucins, monitored as sialic acid,
hexose and absorbance at 280nm, are usually obtained as a single
band at 1.40–1.45 g/ml and may be re–centrifuged under the same
conditions to yield a preparation completely free of non–mucin
proteins and virtually devoid of lipids. It was observed that
decreasing the concentration of guanidinium chloride resulted in
a larger change in the buoyant density of DNA than for the
mucins. Thus, lowering the concentration from 4M to 0.2M
guanidinium chloride increased the density of DNA from 1.48 to
1.70 g/ml whereas the mucins changed from 1.45 to 1.52 g/ml.
Samples containing DNA were thus subjected to a third density-
gradient in 0.2M guanidinium chloride (Figure 3b). This
protocol, designed originally for cervical mucus glycoproteins[21],
works well for both gastric and bronchial mucins. The purified
macro–molecules are denoted 'whole' mucins.

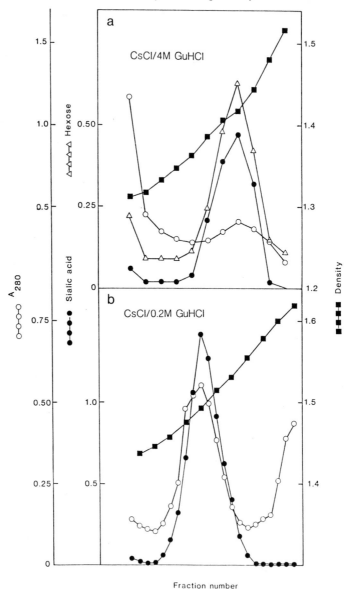

<u>Figure 3</u>   Two-stage isopycnic density-gradient centrifugation of
cervical mucins in (a) CsCl/4M guanidinium chloride and
(b) CsCl/0.2M guanidinium chloride. Reproduced by permission from <u>Biochem. J.</u>, vol. 211, pp. 13–22 copyright
(c) 1983 The Biochemical Society, London.

Mucin Fragments

The whole mucins were cleaved by reduction of disulphide bonds yielding a major fragment (85% by weight) referred to as 'subunits'. The kinetics of this reaction suggested a single population of disulphide bonds and showed that the reaction is complete within 20 minutes. The subunits were recovered as a single band at 1.43 g/ml after isopycnic density–gradient centrifugation in CsCl/4M guanidinium chloride[15].

Trypsin digestion of the subunits yielded a major population of fragments accounting for more than 80% by weight. Kinetic experiments suggested a single population of trypsin–sensitive sites and showed that the reaction is complete within one minute. These fragments, referred to as 'T–domains' to distinguish them from glycopeptides obtained with other proteinases, were also re-covered as a unimodal peak, banding at 1.46 g/ml, after isopycnic density–gradient centrifugation in CsCl/4M guanidinium chloride[15].

3 MACROMOLECULAR CHARACTERISATION

The intact mucins, subunits and T–domains have been studied with a number of physical techniques such as light scattering (both quasi–elastic and absolute–intensity measurements), analytical ultracentrifugation, viscometry, gel chromatography and electron microscopy. The aim of these studies has been to measure molecu-lar mass, size and shape avoiding the effects of self–association and 'aggregation'. The latter consideration is an important one because mucins have a reputation for being 'sticky' and difficult to handle. Most measurements have therefore been performed over a broad range of guanidinium chloride concentrations (0.2M–6M)[22].

Gel Chromatography

Although gel chromatography does not allow a quantitative estimate of M_r, it is a useful tool for studying the size distribution of a population of molecules. We have used this technique for whole mucins (cervical, gastric and bronchial) as well as the cognate subunits and T–domains. In each case we obtain a similar 'fragmentation pattern' which is illustrated here with that for cervical mucins (Figure 4). The whole mucins chromatograph in the void volume and no information about the size distribution is obtained. T–domains elute as a unimodal distribution whilst subunits usually show some evidence of heterogeneity.

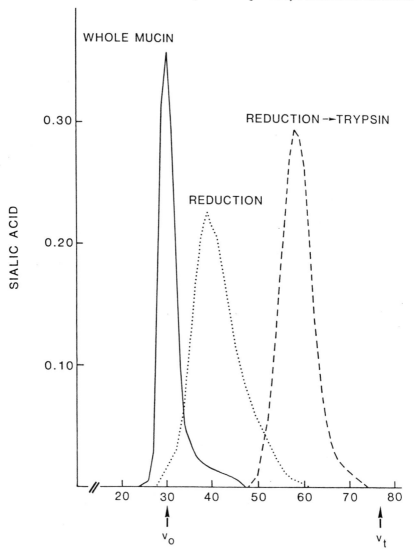

Figure 4   Gel chromatography on Sepharose CL-2B of cervical whole
mucins (——), subunits (······) and T-domains (⤙ – – – ⤚).
The column was eluted with 6M guanidinium chloride and
the fractions analysed for sialic acid. Reprinted by
permission from <u>Biochem. J.</u> vol. 213, pp. 427–435 copyright (c) 1983 The Biochemical Society, London.

Analytical ultracentrifugation

These studies have been performed with an MSE Centriscan 75 analytical ultracentrifuge using schlieren and absorption optics. Three approaches have been used: 1) Isopycnic density–gradient centrifugation to assess the homogeneity and the extent of contamination, particularly with DNA, of the preparations. 2) Sedimentation–velocity to measure sedimentation constants and the concentration–dependence of this parameter. This technique also provides a rapid evaluation of the size homogeneity. 3) Sedimentation–equilibrium to measure M_r of subunits and T–domains, whole mucins being too large for such measurements with the Centriscan.

To obtain M_r from centrifugation data requires that the partial specific volume is known. This parameter has been measured over a range of guanidinium chloride concentrations with a Kratky density meter and is given together with the refractive index increment in Table 1. Sedimentation–velocity measurements were performed as a function of solute concentration and whole mucins, in particular, showed a marked concentration–dependence. When the reciprocal sedimentation coefficient was plotted against concentration (Figure 5) straight lines were obtained up to approximately 4mg/ml. Cervical and bronchial mucins sedimented as single peaks though there was marked broadening at low concentrations indicating polydispersity. A number of gastric mucin preparations also showed some evidence of heterogeneity. Sedimentation constants (s^o) and the cognate concentration-dependence parameters [K_s; defined as $s^o(d(1/s)/dc$] are given in Table 2.

Table 1   Refractive index increment (dn/dc) and partial specific volume ($\bar{v}$) of cervical whole mucins (CPM) and T–domains in guanidinium chloride

GuHCl (M)	dn/dc (ml/g)		$\bar{v}$ (ml/g)	
	CPM	T–domains	CPM	T–domains
0.20	0.161	0.153	0.65	0.65
2.00	0.143	0.137	0.65	0.61
4.00	0.122	0.121	0.66	0.62
6.00	0.110	0.105	0.67	0.65

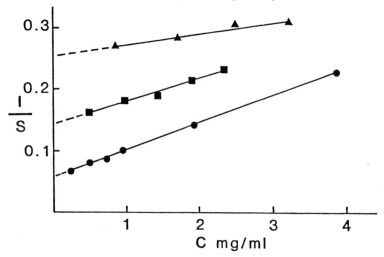

Plot of the reciprocal sedimentation coefficient <u>versus</u>
concentration for cervical whole mucins (●),
subunits (■) and T–domains (▲)

In a treatment of the dependance of K_s on M_r, Rowe[23] has
shown that for large molecules adopting a spherical configuration
in solution

$$M_r = N[6\pi\eta \, s^0/(1-\overline{v}\rho)]^{3/2}/(3K_s/16\pi)^{1/2}$$

where N is Avogadro's number, η the solution viscosity, s^0 the
sedimentation constant and K_s the concentration–dependence parameter of the reciprocal sedimentation coefficient. Values of M_r
obtained for whole mucins with this approach as well as equilibrium measurements (Yphantis' meniscus–depletion method) on subunits and T–domains of cervical mucins are compiled in Table 2.

Table 2 Sedimentation–velocity and equilibrium data for mucins
         and fragments in 6M guanidinium chloride. Subunits and T–
         domains are from cervical mucins (CPM)

Sample	s^0 (S)	K_s (ml/g)	$M(K_s) \times 10^{-6}$	$M(eq) \times 10^{-6}$
CPM	17	760	16	
Subunits	8.5	310	2.9	1.6
T–domains	3.9	83	0.25	0.29
CBM	22	950	23	
PGM	30	890	35	

Light Scattering

Absolute Intensity. These measurements were initially
performed with a Sofica light–scattering photometer at angles
between 30 and 150 degrees and later with a Malvern spectrometer
at 15–150 degrees. Performed over a range of angles and
concentrations, such measurements yield information on the
weight–average M_r and the z–average radius of gyration (R_G)[24].
The technique is sensitive for large molecules and only small
quantities of sample are required; it is non–destructive and
fairly easy to perform over a range of solvent conditions.
However, the presence of large non–representative material such
as 'dust' and 'aggregates' may significantly perturb the results,
and good techniques for handling the samples as well as care in
the interpretation of the data are required. Clarification of
solutions of whole mucins was performed by centrifugation whereas
solutions of subunits and T–domains were also filtered through
0.45μ membranes. Measurements were performed over a range of
guanidinium chloride concentrations to assess whether highly
denaturing conditions would dissolve 'aggregates'. Whole mucins,
unlike T–domains and subunits, could not be effectively
resolubilised after freeze–drying and measurements were therefore
made on dilutions from a stock solution. The accuracy of M_r rely
on the accuracy of the measurement of the refractive–index
increment (dn/dc) and this parameter was determined with either a
Shimadzu refractometer (monochromatic light) or a Waters
differential refractometer (white light). Although the dispersion
of dn/dc with wavelength is about 1–3% we found the Waters
instrument more sensitive and reproducible. Overall we estimate
the accuracy of the light–scattering measurements to be within
20% with good reproducibility. Measurements of dn/dc have only
been performed on cervical 'whole' mucins and fragments and these
values (Table 1) have been used for the gastric and bronchial
mucins.

Table 3   Light–scattering data for mucins and fragments in 6M
guanidinium chloride. Subunits and T–domains are from
cervical mucins

Sample	$M_r \times 10^{-6}$	R_G (nm)	$D_t^o \times 10^8$ (cm^2 sec^{-1})	r_a (nm)
CPM	11	200	1.2	110
Subunits	2.1	74	2.9	46
T–domains	0.38	31	9.3	14
CBM	21	270	0.89	150
PGM	32	350	0.74	180

Zimm plot analysis of absolute–intensity data (Figure 6) allows a rapid assessment of the quality of the data, with the presence of dust/aggregates or errors in a particular concentration of a sample being immediately apparent. Values for M_r for a number of preparations of whole cervical mucins were $10-15 \times 10^6$ and showed no systematic dependence on the concentration of guanidinium chloride though the value for R_G increased slightly between 4M and 6M[22]. Re–solubilised cervical subunits and T–domains did show some evidence of aggregation at low guanidinium chloride concentrations (0–2M) but between 2–6M the values varied only within the errors. Bronchial mucins from patients with chronic bronchitis or cystic fibrosis had values of M_r in the range $15-25 \times 10^6$ for different preparations whereas M_r for gastric mucins varied from about 15×10^6 up to as high as 45×10^6 depending on how they were prepared[16]. The data are summarised in Table 3.

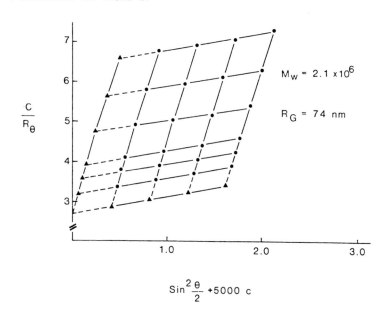

Figure 6 Zimm plot of cervical mucin subunits in 6M guanidinium chloride. Solutions: 0.33, 0.25, 0.16 and 0.08 mg/ml. Measurements were made at 30, 37.5, 45, 60, 75 and 90 degrees. Reproduced by permission from Biochem.J., vol 213, pp. 427–435 copyright (c) The Biochemical Society, London

Quasi—elastic light—scattering. These measurements were made with a Malvern spectrometer and a K7023 digital correlator. The correlation function was fitted to a second order polynomial by least—squares analysis and the translational diffusion coefficient (D_t) calculated from the linear term of the quadratic fit. Procedures for solution clarification were the same as those used for absolute—intensity work. Measurements were performed as a function of angle and concentration and the values of D_t extrapolated to zero. Values obtained for a variety of mucins and fragments together with the calculated Stokes radii (r_a) for the equivalent spheres are given in Table 3.

Viscometry

This technique has mainly been used on cervical whole mucins, subunits and T—domains[18]. Measurements were performed with a modified cone and plate viscometer over shear rates in the range 0.01—100 sec^{-1} at concentrations from 0.5 to 4mg/ml for whole mucins and up to 20mg/ml for T—domains. The whole mucins, in particular, showed a marked shear—dependence and values for the intrinsic viscosity (extrapolated to zero shear) of 635ml/g (whole mucins), 224ml/g (subunits) and 65ml/g (T—domains) were obtained.

Rate—zonal Centrifugation

As mentioned previously, no precise information is obtained about the distribution of M_r for whole mucins either from gel chromatography or conventional sedimentation—velocity in the analytical ultracentrifuge. Creeth and Cooper[25], using very low-speed equilibrium runs performed on a Beckman Model E centrifuge, have developed a procedure by which data are analysed to obtain distributions of M_r for molecules up to approximately 10×10^6. As an alternative approach, we have used rate—zonal centrifugation on preformed gradients of either guanidinium chloride or sucrose[26]. Guanidinium chloride was used for quantitative work because sugar analysis is easier to perform in this solvent whilst sucrose gradients could only be monitored by UV absorption at 280 and 206nm. Samples were layered onto preformed gradients (6 to 8M guanidinium chloride) and run for 2.5h at 45000 revs/min in a 6x5.2ml swing—out rotor. The gradient was emptied from the bottom and D_t, M_r and R_G measured on the individual fractions with light—scattering methods (Figure 7).

The data showed that the mucins are very disperse in size with a broad distribution of M_r. Possibly, three sub—populations centered around 6×10^6, 16×10^6, and 24×10^6 are present. The average value of M_r (11×10^6) is in good agreement with those obtained for the unfractionated material by light—scattering methods (Table 3).

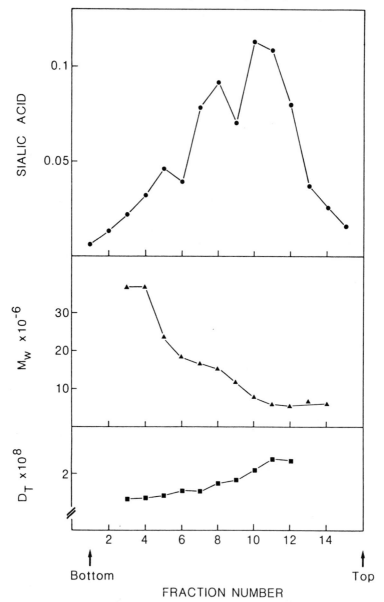

<u>Figure 7</u>   Rate–zonal centrifugation of cervical mucins in a
             guanidinium chloride gradient. Reproduced by permission
             from <u>Biochem. J.</u>, vol. 245, pp. 757–762 copyright (c)
             1987 The Biochemical Society, London

Hydrodynamic Models

The growth of the volume domain of a molecule (as measured by parameters such as $[\eta]$, s^o, R_G or D_t^o) with M_r provides information about the likely molecular model[24]. A prerequisite for this approach is that the molecules are studied with cognate fragments and/or have been fractionated into sub-populations. A plot of $\log R_G$ versus $\log M_r$ for various mucins and fragments yields a slope (α =0.5–0.6) which is consistent with a random coil rather than with a rod-like ($\alpha > 1$), a heavily branched or a star-like structure ($\alpha < 0.4$). The fact that the values for T-domains and subunits appear on the same line as those for the whole mucins suggests that these fragments share the properties of the intact macromolecules. When values for the intrinsic viscosity, D_t^o and s^o are available, the relative molecular mass may be determined with the Sheraga–Mandelkern and the Svedberg equations (Table 4). These values are in agreement with those obtained with total-intensity light-scattering.

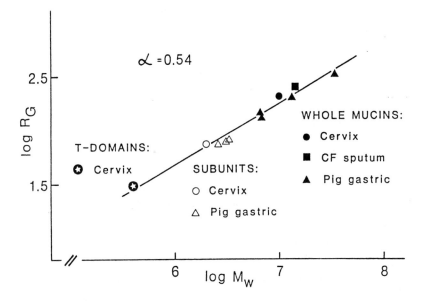

Figure 8   Plot of $\log R_G$ versus M_r. The value for the slope (0.54) suggests a random coil structure for the molecules. Reprinted by permission from Biochem. Soc. Trans., vol 12, pp. 615–617 copyright (c) 1984 The Biochemical Society, London

Table 4   Values of M_r calculated from hydrodynamic and light-
scattering data

Sample	s^o/D_t^o $\times 10^{-6}$	$s^o/[\eta]$ $\times 10^{-6}$
CPM	15	12
Subunits	2.9	2.2
T-domains	0.40	0.35
CBM	25	
PGM	46	

Electron Microscopy

Hydrodynamic data alone cannot usually provide conclusive
evidence for a particular structural model for large and
complicated macromolecules and we have therefore used electron
microscopy to gain further insight into the 'architecture' of the
mucus glycoproteins. The molecules were spread in a monolayer of
benzyldimethylalkyl ammonium chloride (BAC)[27] and visualized by
using rotary shadowing and/or staining with uranyl acetate[19]. In
Figure 9, some typical results for cervical whole mucins,
subunits and T-domains are shown. The particles are clearly
linear and may adopt very different 'shapes'. The latter feature
is interpreted as the macromolecules having a pronounced
configurational freedom, i.e. they are highly flexible.

4 DISCUSSION

Size of whole mucins

When precautions are taken against mechanical and proteolytic
cleavage during extraction and purification, mucus glycoproteins
are undoubtedly very large macromolecules (M_r 10–40x10[6]) which
are polydisperse and possibly heterogeneous in a number of
physical properties such as size and buoyant density. Measurements of M_r on disperse distributions of large molecules are
difficult to perform and both the principle 'absolute' methods
available (light–scattering and equilibrium ultracentrifugation)
suffer from short–comings which tend to introduce systematic
errors.

Total–intensity light–scattering is considered to give
'high' values of M_r and R_G owing to difficulties connected with
removing residual large aggregates of the sample and 'dust'. On
the other hand, there is a possibility of a selective loss of the
larger species of a polydisperse distribution during clarification by procedures such as filtering and high–speed centrifu-

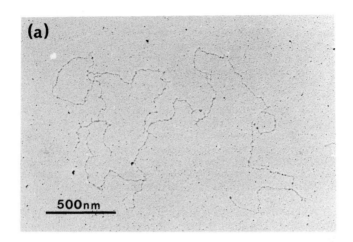

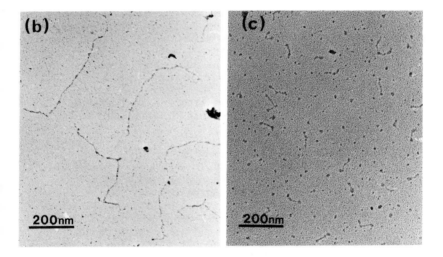

Figure 9.   Electron microscopy of cervical whole mucins (a),
subunits (b) and T-domains (c). The whole mucins
which are filamentous and flexible range in size
from 500–5000nm with a number average length (ln)
of 1240nm. Subunits with lengths from 200–600nm

gation. No information on the size-distribution is obtained with this technique. Quasi-elastic light-scattering performed, for example, as photon correlation spectroscopy is also perturbed by the presence of large non-representative particles but the contribution to the final result is more difficut to evaluate. Large particles may act as local scatterers introducing 'partial heterodyning' in a homodyne experiment as well as raising the 'base-line' of the correllogram. The latter effect may to some extent be avoided by data exclusion algorithms. Enough 'channels' must be available in the correlator to 'accomodate' the whole distribution and thus ensure that the smaller species are not excluded. The techniques available for evaluating the size-distribution with quasi-elastic light-scattering suffer from too many assumptions concerning the shapes of the molecules to be of any real value here. Furthermore, since the relationship between the intensity of the scattered light and the size of the molecules is not known, distributions of D_t cannot be expressed on a weight basis. In short, light-scattering techniques will tend to over-estimate the size of large molecules and this effect may be exaggerated when data are extrapolated to zero angle where the contribution from the larger species is most pronounced.

Analytical ultracentrifugation (low-speed equilibrium techniques) on the other hand, may underestimate the average M_r of a polydisperse distribution owing to losses of the largest species at the cell bottom. Furthermore, it is not clear whether there is a systematic variation of the partial specific volume with size and, if so, how this would influence the result. The technique is, however, very sensitive and samples can be studied at high dilutions. Distributions of M_r may be obtained and the only major draw-back is that, even with the best available data-handling routines, molecules larger than M_r about 10×10^6 will be excluded from the analysis.

In our opinion, the best strategy is to use a number of techniques for the determination of M_r and try to reach the truth in a crabwise fashion. Hydrodynamic parameters such as s^0 and $[\eta]$, which are less perturbed by aggregates and 'dust', may be used to evaluate how realistic the values are. We would advise investigators not to rely on a single technique when studying the macromolecular properties of large and complicated structures such as mucins. It should be pointed out that it is not possible from physical data to conclude that the macromolecules are covalent entities although measurements were performed at very low concentrations and over a wide concentration range (0.2–8M) of guanidinium chloride, a highly dissociating agent.

Size of mucin fragments

Subunits obtained after reduction of mucins (cervical, bronchial and respiratory) isolated with the procedure outlined above are significantly larger (M_r about $2-3\times10^6$) than usually obtained by most investigators (500 000). Measurements of M_r are much easier to perform in this size–range and it is clear that the discrepancy cannot be explained by technial aspects. It could be argued that all disulphide bonds may not have been cleaved but kinetic experiments, as well as the fact that the conditions we used for reduction are at least as 'severe' as those used by others, speak against this explanation. We do obtain subunits with M_r approximately 500 000 if the cognate whole mucins are first subjected to a very brief trypsin digestion and this shows that the size of subunits may reflect the extent to which the cognate whole mucins have been subjected to proteolysis prior to reduction. However, this does not prove that the protein core in the 'large' subunit is a continous polypeptide. Trypsin digestion of subunits affords high–M_r glycopeptides (300 000–400 000), i.e. of similar size as the 'basic unit' defined by other investigators. A subunit thus contains a number (4–5) of 'basic units' (heavily glycosylated domains) on the same protein core which is folded and stabilized by disulphide bonds between the glycosylated domains. The protein core of the subunits may be the primary translation product but that of the glycopeptides is clearly not. These units are obtained after scission of peptide bonds. We would therefore advocate that a subunit, with the size and structure outlined above, is the basic building block of mucins and, consequently, must be accomodated in any model proposed for their macromolecular structure.

Structural Models

Hydrodynamic data on its own cannot be used to 'prove' a molecular model but, if accurate, must be consistent with the correct structure. The 'windmill' model can neither account for the large subunit described above nor explain the large size and broad size distribution of the whole mucins and the discrepancy between M_r 2×10^6 and $10-40\times10^6$ is too great to leave unaccounted for. Placing more subunits around the central linking protein would be inconsistent with the hydrodynamic data relating size (R_G, D_t^o, $[\eta]$, s^o) to M_r and, more important, we find no evidence for a regular 'star' structure by using electron microscopy. However, we do recognize the conceptual impact of this model. In particular, it was the first to focus on mucins being formed from smaller units joined by disulphide bonds.

The 'lectin' model could, in principle, explain the large size, the polydispersity, the hydrodynamic data and the linear structures observed in the electron microscope. However, two observations make the model less likely: 1, The size of 'our' subunit compared to the 'basic unit' upon which this model is based and 2, the fact that guanidinium chloride (up 8M) has no effect on M_r for the whole mucins and the fragments. In the absence of evidence for a reversible change in M_r, by variation of solvent conditions or inhibition at the lectin site with specific carbohydrate structures, the model must be regarded as an interesting speculation.

The 'swollen coil array' model is similar to the 'linear random coil' model in that a linear and flexible molecule adopting a coiled conformation is postulated for the 'basic units' (T–domains) and the overall macromolecule. However, no proposals as to the nature of the mucin segments linking the basic units is given, other than it is more extended and less glycosylated. The slow rate of degradation of whole mucins, as opposed to subunits, by trypsin is not consistent with these extended regions being sensitive to proteolysis and recent electron microscopy evidence using colloidal gold as a probe for 'naked' protein regions suggests they are highly localised (Sheehan and Carlstedt unpublished)

In summary, cervical, gastric and bronchial mucins are very large (M_r=10–40x106) macromolecules composed of subunits (M_r 1–3x106). Trypsin digestion of the latter yields high–M_r glycopeptides (300 000–500 000) similar in size to the 'basic' unit described by others. Subunits are regarded as the 'basic' unit rather than the latter because fragments of this size are only obtained via the action of proteinases, either on subunits (step–wise degradation) or on the whole mucins (proteolytic modification). Hydrodynamic data and electron microscopy of the whole mucins and the fragments favour a linear random coil model for the macromolecules.

ACKNOWLEDGEMENTS

This work was supported by the Cystic Fibrosis Research Trus UK, the Wellcome Trust, UK, the Swedish Medical Research Council (project no. 7902), the National Swedish Board for Technical Developments and the Medical Faculty of Lund, Sweden. We thank Joy Greenwood for making the schematic drawing and David, J. Thornton for valuable suggestions.

REFERENCES

1.   I. Carlstedt, J.K. Sheehan, A.P. Corfield and J.T. Gallagher, Essays Biochem., 1985, 20, 40.
2.   D. Snary, A. Allen and R.H. Pain, Biochem. Biophys. Res. Commun., 1970, 40, 844.

3. B.J. Starkey, D. Snary and A. Allen, Biochem. J., 1974, 141, 633.

4. D. Snary, A. Allen and R.H. Pain, Biochem. J., 1974, 141, 641.

5. A. Allen, R.H. Pain and T.R. Robson, Nature (London), 1976, 264, 88.

6. M. Scawen and A. Allen, Biochem. J., 1977, 163, 363.

7. J.P. Pearson, A. Allen and S. Parry, Biochem. J., 1981, 197, 155.

8. A. Allen, Trends Biochem. Sci., 1983, 8, 169.

9. A. Allen, D.A. Hutton, D. Mantle and R.H. Pain, Biochem. Soc. Trans., 1984, 12, 612.

10. A. Allen, D.A. Hutton, J.P Pearson and L.A. Sellers, Ciba Found. Symp., 1984, 109, 137.

11. S.E. Harding, J.M. Creeth and A.J. Rowe, 'Proc. 7th Int. Symp. on Glycoconjugates', Rahms, Lund, Sweden, 1983, 558.

12. S.E. Harding, A.J. Rowe and J.M. Creeth, Biochem. J., 1983, 209, 893.

13. P. Hallet, A.J. Rowe and S.E. Harding, Biochem. Soc. Trans., 1984, 12, 878.

14. A. Silberberg, Biorheology, 1987, 24, 605.

15. I. Carlstedt, H. Lindgren and J.K. Sheehan, Biochem. J., 1983, 213, 427.

16. I. Carlstedt and J.K. Sheehan, Biochem. Soc. Trans., 1984, 12, 615.

17. I. Carlstedt and J.K. Sheehan, Ciba Found. Symp., 1984, 109, 157.

18. J.K. Sheehan and I. Carlstedt, Biochem. J., 1984, 217, 93.

19. J.K. Sheehan, K. Oates and I. Carlstedt, Biochem. J., 1986, 239, 147.

20. S.W. Sajdera and V.C. Hascall, J. Biol. Chem., 1969, 244, 77.

21. I. Carlstedt, H. Lindgren, J.K. Sheehan, U. Ulmsten and L. Wingerup, Biochem. J., 1983, 211, 13.

22. J.K. Sheehan and I. Carlstedt, Biochem. J., 1984, 221, 499.

23. A.J. Rowe, Biopolymers, 1977, 16, 2595.

24. C. Tanford, 'Physical Chemistry of Macromolecules', Wiley & Sons, New York, 1961.

25. J.M. Creeth and B. Cooper, Biochem. Soc. Trans., 1984, 12, 618.

26. J.K. Sheehan and I. Carlstedt, Biochem. J., 1987, 245, 757.

27. T. Koller, A.G. Harford, Y.K. Lee and M. Beer, Micron, 1969, 1, 110.

18

Conformations and Dynamics of Saccharides

By R. J. P. Williams[1] and D. L. Fernandes[2]

[1]DEPARTMENT OF INORGANIC CHEMISTRY, OXFORD UNIVERSITY, OXFORD, U.K.
[2]THE GLYCOBIOLOGY UNIT, DEPARTMENT OF BIOCHEMISTRY, OXFORD UNIVERSITY, OXFORD, U.K.

1   INTRODUCTION

In most organisms the outside (exo– and ecto–cellular) regions of the cell have carbohydrate associated with proteins (i.e. glycoproteins and proteoglycans) and lipids (i.e. glycolipids) or in hydrated polysaccharide gels. In unicellular organisms the combination of these macromolecules, peptides and saccharides, gives rise to cell walls, protective polymers, receptors, and enzymes both in membranes and in the periplasmic and extra–cellular spaces. In multicellular organisms, while the same diversity exists, the polymers also form the structural network which holds the organisation together. We shall discuss the structures and dynamics of large polysaccharides and the saccharides of glycoconjugates with special reference to their membrane associated functions. For reviews on saccharide structures and functions see references 1–8.

Most serum–derived and cell–surface proteins are deliberately glycosylated, i.e. have covalently attached oligosaccharides.[9] Although these 'glyco' parts have been clearly implicated in a number of biological phenomena, their precise functional significance in many cases remains obscure. The primary structure of these highly branched oligosaccharides is often very complex since their constituent monosaccharides can be linked in many different ways. Consequently, the potential information encoded into an oligosaccharide via its primary sequence and three–dimensional structure is considerable. In addition to these oligosaccharides, which are frequently of relatively short chain length, there is a huge variety of very long chain polysaccharides, present in connective tissue for example, which have a diversity of structural functions. We are interested here in a comparison between the structures of all these different polymers relative to their functions.

The aim of this article is not to comment on the nature of the saccharide–associated proteins or lipids to any great degree but rather to pose questions about the nature of the saccharides as polymers in an attempt to relate their functional significance to their composition *via* their structures and the dynamics of these structures. However, it will be necessary to start from some comments concerning the general and familiar nature of other biological polymers.

2 BACKBONES OF BIOLOGICAL POLYMERS

Proteins

The backbone of proteins is formed from the link $-[-RCH-CO-NH]-$. The peptide group $-CO-NH-$ is relatively rigidly fixed in a plane so that protein dynamics and folding concern the flexibility about the unit $-RCH-$. There is one exceptional amino–acid, proline, of greatly reduced flexibility and which has a major role in the control of folding.

Nucleic Acids

The backbone of polynucleotides consist either of 3',5'–ribose phosphate or its 2–deoxy analogue. Here, the unit has much higher flexibility in that there is effectively free rotation at C–5' and at each of the ester oxygens of the phosphorous as well as conformational mobility within the 5–membered ribose ring since various endo/exo forms easily interconvert.

Polysaccharides and Oligosaccharides

Saccharide backbones are generally based on the condensation of five and rigid six–membered cyclic oxide rings (furanose and pyranose respectively). Here different linkages are possible, unlike proteins and DNA where all the linkages in a molecule are of one type. Note that the anomeric carbon (C1 for most monosaccharides, C2 for sialic acids) is involved in the linkages (e.g. 1→2, 1→3, 1→4, or 1→6 links). This fixes the position at this site of the oxygen atom with respect to the ring. If this were not the case, the ring would open at the anomeric position to yield a set of isomers with the OH group directed either axially or equatorially with free interchange. This would degrade the unit as a recognition site. As the anomeric oxygen may lie axial or equatorial to the plane of the ring, there arises the possibility of two isomers for each monosaccharide, an α–form (O–axial to the ring) and a β–form (O–equatorial to the ring). Note that, in some sense, a large ring such as a monosaccharide can be regarded as a short inserted double chain.

In the case of glycosidic (i.e. ether) links from any of the oxygens 1, 2, 3, 4, and 6 and any other such oxygen of another monomer there is flexibility around the C–O–C unit but this is limited by the steric properties of the carbon atom links to two other carbon atoms. There is greater flexibility *via* the 6 linkages since here the ether bond is in the structure (hexose ring)–CH_2–O–(hexose ring), and this not only introduces rotamers of its own, but also increases the freedom around the ether oxygen. These points are well illustrated in the two dimensional energy contour maps of Figure 1 which are the result of a series of *in vacuo* potential energy calculations performed using an AM1 hamiltonian. (This hamiltonian is a parameterised representation of the Shroedinger equation for a multi–atom system).[10] The energy was calculated at each of 144 points on the surface, representing rotations of 360° in 12 steps about the ϕ and ψ bond angles of the glycosidic links. The contour plots show that the range over which ϕ and ψ vary in the case of the 1→2 and 1→3 linkages is very much more restricted than in the 1→6 case. These results have also been verified experimentally using NMR methods.[10-14]

From the above, one might expect that chain flexibility would follow the order:

single DNA and RNA > 1→6 sugar > proteins > 1→(2,3,4) sugars

with the DNA and RNA molecules having the greatest flexibility for a single chain due to the phosphate linkages. However, we must also observe that in proteins every third chain atom is a point of considerable mobility, while in the DNA, RNA, and polysaccharides mobility at two or three consecutive atoms is followed by a relatively large and rigid saccharide unit. Finally, there is the effect upon conformation of the α or β configuration at C–1. Generally, β1→4 structures are associated with more rigid elements than α1→4 structures and, interestingly, one finds that the core regions of complex class oligosaccharides are composed of β1→4 linked residues.

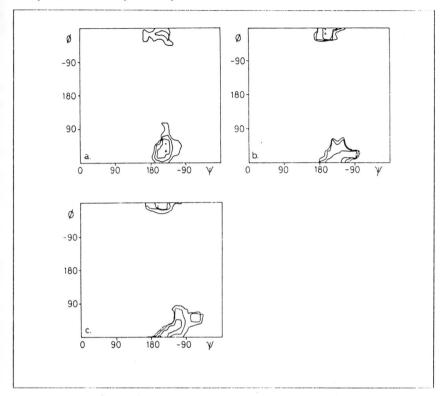

Figure 1 : Energy Maps for Saccharide Units

Potential surfaces for the Manα1→3Manβ linkage in (a) Manα1→3Manβ1→4GlcNAcβ (b) Manα1→3(GlcNAcβ1→4)Manβ, and (c) Manα1→3(Xylβ1→2)Manβ calculated using MNDO procedures. Each surface is contoured at the energy bounds for 70%, 95%, and 99% of the molecules. The experimental and theoretical minimum energy values are denoted Δ and + respectively. Note the restriction on the range of the angles. (After reference 12).

3   SIDE CHAINS : PHYSICAL PROPERTIES

Proteins

The side chains of proteins are characterised by a variety in physical and chemical properties from some 20 amino acids with one common flexibility characteristic apparent. Almost all are based on $\alpha CH–CH_2–X$ units. No doubt, the construction of most amino acids with a β –CH_2 is to provide mobility as well as extension. The X can provide very water soluble units, when the protein tends to be extended in water, or very water insoluble groups when it tends to fold. The huge variety of X in an n–mer peptide (i.e. a protein) generates the rich variety of the fold in lipid or in water. This variety also permits every kind of strength and mobility in the fold since X can be more or less hydrophobic on average. Overwhelmingly, proteins are folded structures of relatively low mobility.

Nucleic Acids

In contrast to proteins, the side chains of the polynucleic acids are few and very similar. They have no flexibility, being aromatic ring systems which can only rotate with much restriction about sugar–base bonds and are to a fair degree equally hydrophobic. All five are found attached to the hydrophilic pentose phosphate polymer. Consequently, there is one type of basic fold, the double helix, with bends and crucifix structures now and then. This is the only basic quaternary fold or structure. Note that diversity both in proteins and nucleic acids is achieved by unique *linear* coupling in sequences.

Saccharides

Saccharides are quite variable in side–chain structure.[1,8] Generally, the side chains help to make the polymer more water soluble. Moreover, they are not so mobile as proteins in the sense that they are not connected *via* –CH_2– units to the main chain, the single exception being the –CH_2OH unit attached to carbon 5 in pyranoses. Of the two faces to each saccharide ring, one is generally more hydrophilic than the other, due to the greater number of hydroxyl groups on one side of the ring than the other. This is especially marked in monosaccharides such as mannose. The hydrophobicity of a polysaccharide is then based on *one face* of the hexose ring, i.e. a property of the backbone and not in the side chains, in marked contrast with those of proteins and polynucleic acids. On the whole, the fold of a polysaccharide is based on main–chain – main–chain inter-action, as in a protein helix, and tertiary structures are not strong. It is the weak-ness of the tertiary and quaternary structure of many saccharides (but not all) that make them a very special set of biopolymers, sometimes lacking structural defini-tion and therefore having very different roles from proteins and polynucleotides. Above all, it is the interaction with solvent water which dominates their conformations. Note that diversity lies now in sequence and branching and in the way in which structures can be combined within populations. Recently, studies of a number of glycoconjugate systems have suggested that such diversity in glycoprotein oligosaccharides creates discrete subsets (so—called glycoforms) that have different physical and biochemical properties and that these may reflect functional diversity.[9]

At this point it is instructive to consider the variety of polymers and oligomers that can be built from the simple building blocks.

4   THE INFORMATION CONTENT OF BIOPOLYMERS

The relative complexities of the various biopolymers can be gauged by considering the number of different structures of a given length that could occur. For proteins, there are 20^n ways of assembling a polypeptide of length n from twenty amino acids. For example, there are 8000 possible tri—peptides and 1.28×10^{13} possible decapeptides. Similarly, nucleic acids with four different monomer types can form 4^n polynucleotides of length n. The situation is more complex for saccharides since the individual monomers can be linked in different ways. In this case, the number of ways of assembling a linear (i.e. non—branching) saccharide of length n from θ different monosaccharide types (e.g. galactose, mannose, N—acetyl glucosamine, etc.), each of which can exist in α or β forms, and which can be linked to an adjacent residue in one of ξ different ways (e.g. 1→2, 1→3, 1→4, etc.), is given by

$$N_{sacch(n)} = \theta(2\theta\xi)^{n-1}$$

This value is doubled if the anomeric configuration of the reducing terminus residue is taken into account.[15] Examples of the number of possible linear sequences for the three types of biopolymer are given in Table 1. It is very probable that some of the saccharides could not physically occur, for example through steric constraints. Even so, it is clear that the inherent complexity of oligosaccharides is much greater than that of the other types of biopolymer and that even short sequences have high theoretical information content. If one considers branched structures the number of possible sequences is enormous. Note that one talks of theoretical information content and there is no suggestion that information is encoded directly in the primary structure of saccharides, as it is with nucleic acids. Rather, the large number of possible sugars indicates the

Table 1 : Number of Linear Oligomers of Length N for various Biopolymers

N	DNA	Proteins	Oligosaccharides* $\Theta = 4$	$\Theta = 8$
1	4	20	4	10
2	16	400	128	800
3	64	8 000	4 096	64 000
4	256	160 000	131 072	2 097 152
5	1 024	3 200 000	4 194 304	1.34×10^8
6	4 096	6.40×10^7	1.34×10^8	3.27×10^{10}
7	16 384	1.28×10^9	4.29×10^9	5.49×10^{11}
8	65 536	2.56×10^{10}	1.37×10^{11}	3.52×10^{13}
9	262 144	5.12×10^{11}	4.39×10^{12}	2.25×10^{15}
10	1.04×10^6	1.28×10^{13}	1.40×10^{14}	1.34×10^{18}

*Θ = The number of monosaccharide types (see text). The table shows the number of ways of constructing a linear oligomer of length N for the three main types of biopolymer. The figures for the oligosaccharides are based on the assumption that each monosaccharide type can exist in either α or β anomer forms and can participate in one of four different types of linkage (e.g. $1\rightarrow2$, $1\rightarrow3$, $1\rightarrow4$, or $1\rightarrow6$). Note how rapidly the diversity of potential oligosaccharides increases with chain length. For example, the number of ways of assembling a non−branching deca−saccharide from eight different monosaccharide types is over one hundred thousand times greater than that for constructing a decapeptide from twenty amino acids. Note also that branching greatly increases the number of theoretical possibilities but steric constraints would limit the number of physically possible structures.

wealth of opportunity for constructing saccharides with different shapes. This is similar to proteins with their enormous diversity in structure and function. However, in the case of saccharides, branching and heterogeneity in linkage allow tremendous diversity with a small number of monomers, giving relatively well–defined surfaces *without extensive folding*.

5 SPACE FILLING

Proteins

The ability of a linear polymer to fill space depends on the flexibility of the joints and the sizes of the units. It is noteable that proteins are polymers of small peptide units. For this reason, a complete turn can be made in a small space so that the helices, β–sheets, and β–turns, assisted by H–bonds, can form and fill space easily in very effective ways, leaving surfaces to generate many properties and little or no room for water in the interior. Their folds are admirably suited to provide

i   almost rigid platforms for recognition, e.g. enzyme active sites and receptors mostly based on β–sheets,

ii   stable rod–like elements of considerable length based on α–helices, e.g. the mechanical transmission rods of biology,

iii local energised regions generated by the strength of the whole fold, and

iv   effectively, a 3–dimensional form of information from a relatively rigid folded linear sequence.

Often, concave as well as convex surfaces are formed.

Polynucleotides

Polynucleotides are obviously based on quite another principle in that the unit is an anion. A polyribosephosphate will resist tertiary folding but, as with proteins, the double helix interior of the bases of DNA and RNA in dimeric units contains little or no water. DNA and RNA folding is controlled by side–chains or added cations of various forms, e.g. Mg^{2+}, polyamines, and histones in the grooves. The consequences of the requirements for side–chain H–bonding and for the hydration of the soluble ribose phosphate are such that by themselves these polymers tend to form long strings. This is exactly what is required if they are to be self replicated, transcribed, and translated. Their space filling is not restricted by the character of their backbone flexibility, however, but is controlled by their double stranded nature. They are a linear form of information only since, in large part, they have few of the properties given for proteins above.

Saccharides

Glycans have an interesting position relative to these other two classes of polymers since, even as unbranched polymers, they have rather a large rigid single unit, the hexose ring, which in polymeric form tends to give an extended chain due to its considerable hydrophilicity. A likely secondary structure open to the backbone is the helix and there appears to be no way in which tertiary fold would then be generated, since the exterior of these helices are very hydrophilic (see for example starch).[16] Their shape is bound to be open to contact with water. However, three factors suggest that we should not be looking for simple principles concerning the fold of these molecules. They are :

1   chain branching,

2   alternating chain linking (i.e. they can be hetero–polymers within the backbone), and

3   the curiousness of their side chains.[1,3,8]

These molecules are such that a great variety of types of polymer can be constructed. While the above limitations on these constructions makes them of reduced value in, for example, the generation of recognition surfaces by folding as found in proteins, they become a form of information through variations in sequence, connectivity, and branching. These together provide quite new ways of generating information in three dimensions i.e. without a rigid fold. Some polysaccharides can adopt rigid folds but they are of limited sequential information content. In fact, it appears that to generate specific folds, polysaccharides must be built from simple A_n or $(AB)_n$ type units. A specific fold of a wider diversity of units needs to be synthesised free from errors on a template and polysaccharides are not built in this way. Regarding their potential functions, perhaps for the very reason that they are limited in their folding they do not have local regions which generate enzyme *activity*. It is clear that the great advantage of the saccharide polymers is that they can build *passive shapes* of either great flexibility or rigidity outside cells and often connected to cell membranes.

6 CHAIN BRANCHING

Some simple points should be clarified. Firstly, chain branching can provide either a flexible or a relatively inflexible connection (e.g. 1→6 and 1→3 linkages, respectively). A common branching pattern in glycoproteins is shown in Figure 2. An important question is the significance of rigidity and mobility at branching points. In this regard, one can consider the forms of trees and umbrellas. Branching after starting from a linear trunk provides the most economical way of generating a large surface while covering the ground from which the trunk emerges. Note that a branched saccharide can

i   protect a large part of the surface of a protein or a membrane due to the size of the monomer,

ii provide a large exposed surface, especially on a membrane carrying recognisable features using few residues, and

iii bear functional units without defined folding based on a sequence.

Furthermore,

iv branching in a group of large polymers is effectively a cross-link which reduces freedom of movement and creates entanglements.

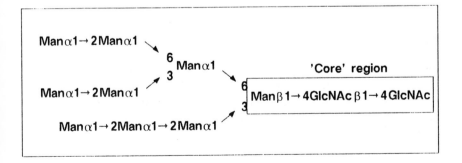

Figure 2 : Sequence of N−linked Oligosaccharide from Soybean Agglutinin

7   SIDE CHAINS : CHEMICAL NATURE

The most common saccharide side chains are the fixed −OH groups, the 6 CH_2OH, and the $NHCOCH_3$ group. These units are obviously highly effective H−bonding centres and allow solvation as well as intra− and extra− chain−chain interactions. Other common units are anions $-CO_2^-$ and $-OSO_3^-$ which would repel one another. These units will be discussed later. The few hydrophobic units include O−methyl groups. Now the distribution of the side chains is unique to a given saccharide and one can ask what properties are conferred by, for example, sulphation. It must also be remembered that unlike proteins the side chains can be attached at several points around the hexose ring so that once again three−dimensional information can be developed without folding so that a scheme like Table 1 can be developed for each side chain as well as for chain connectivity.

A special feature of the $NHCOCH_3$ group of the saccharides is that it permits the formation of secondary structure using side−chain H−bonding. Some of the most rigid saccharides are based on the β−sheet formed from the antiparallel strands of [NH−CO] units which can be compared to the β−sheet of proteins.[17] The β−sheet and the α−helix are perhaps equally rigid but the (curved) sheet effectively forms a strong three−dimensional rigid surface while the helix is effectively a one−dimensional rod with flexibility between rods.[18] These differences can be expected to reflect differences in the functions of these structures.

8   FUNCTIONS OF BIOLOGICAL POLYMERS AND
OLIGOMERS

The major functions of biological polymers and oligomers can be classified as in Table 2.

Regarding saccharides, their possible functions fall into three broad categories (1) energy storage, (2) structural roles (including functions based on mobile as well as rigid structures), and (3) involvement in recognition phenomena. We are interested in the latter two and, specifically, the relation between saccharide form and function.

Table 2 : Major Functions of Biological Polymers

Function	Saccharides	Other Biopolymers
Stores of genetic information	no	DNA, RNA
Catalytic surfaces	none known	protein/enzymes
Structural supports	cellulose	collagen, silk
Recognition devices	oligosaccharides	antibodies
Physical (Mechanical) devices	proteoglycans	muscle proteins
Protective devices	lignins	coat proteins
Energy storage	glycogen	seed proteins

Structural Roles of Saccharides

The roles of homo– and hetero–polysaccharides as structural elements in cells are well known and extensively reviewed elsewhere.[1,4,8] These structures include starch, cellulose, gums, mucilages, pectins, hemicelluloses (in plants), teichoic acids, cell wall peptidoglycans, and extracellular and capsular polysaccharides in bacteria, chitin (invertebrates), and fungal and algal cell wall saccharides.

Amongst these polysaccharides there are those that have an ability to have a rigid three–dimensional structure such as cellulose but there is also another group of polysaccharides which have virtually no defined 3–D structure, and these mobile molecules (e.g. mucilages) have quite different purposes associated with their dynamics. We return to the discussion of the dynamics of the molecules in a later section.

Turning to other structural roles, carbohydrates are known to influence the chemical and physical properties of glycoproteins.[19,20] These properties include three–dimensional structure and susceptibility to degradative enzymes and heat. For example, the activity of 'antifreeze' glycoproteins, which inhibit the formation of ice crystals in the blood of arctic fish, is destroyed upon modification of its carbohydrates.[17]

For many glycoproteins, removal or alteration of the carbohydrates results in no qualitative change in functional capability but rather a change in physical stability. For example, in some glycoenzymes, removal of carbohydrates causes no change in catalytic activity but does result in decreased heat stability. Numerous aglycosyl glycoproteins, (for example fibroblast fibronectin, the acetylcholine receptor of embryonic muscle, and muscle fusion proteins produced by tunicamycin treated cells), show increased susceptibility to proteolysis compared to the intact glycoproteins.[20] This has also been found in enzymatically deglycosylated glycoproteins. Aglycosylated monoclonal IgG produced in the presence of tunicamycin (an inhibitor of glycosylation) retains normal antigen, protein A, and C1q binding, and C1 activation but shows a 60-fold increase in the rate of proteolysis by pepsin, loss of binding to monocyte Fc receptors, and loss of the ability to induce antibody–dependent cellular cytotoxicity and, in complexes with antigen, failed to be cleared rapidly from the circulation.[21] The current theory is that carbohydrates confer protection against proteolysis by concealing potential cleavage sites, for example in fibronectin and the haemaglutinin of influenza virus. We have already seen the advantages of using branched saccharides for this role. A masking function has also been proposed for the variant surface glycoproteins (VSGs) which form a dense cell surface coat in the malaria parasite. The VSG coat seems to act as a permeability barrier to macromolecules, so protecting the organism from attack by the host's immune system. [22]

Carbohydrates may also induce conformational changes in glycoproteins. This has been shown in fibronectin (a molecule responsible for anchoring and guiding the migration of cells), the three dimensional structure of which is altered by both heparin and heparin sulphate (the carbohydrate to which it normally binds). This saccharide–induced change in three–dimensional structure is thought to play a role in maintaining the integrity of the molecule's cell binding domain.[23]

Recognition Phenomena

The cell surfaces of multicellular organisms are equipped with receptors and ligands for the recognition of free molecules, specific substrates, and the surfaces of other cells.[2,24] Such machinery is also found within cells and is involved, for example, in the intracellular routing of proteins. Numerous studies have suggested that saccharides (specifically branched oligosaccharides) participate in such recognition events.[24-28] These phenomena can be classified as cell–molecule, cell–substratum, and cell–cell interactions, and intracellular routing of glycoproteins. Other phenomena of importance to saccharide function include species– and organ–specific glycosylation, microheterogeneity, and site–specific glycosylation. These are discussed in more detail below.

Cell–Molecule Interactions Carbohydrate ligands are required in many cell–molecule interactions. For example, cell surface oligosaccharide receptors are involved in the clearance and segregation of humoral glycoproteins by hepatocytes, endocytosis in macrophages, and ligand internalization. The liver cell receptors have been particularly well studied, and include those specific for mannosyl or N–acetylglucosaminyl, fucosyl, galactosyl, and phosphomannosyl residues.[29]

Carbohydrates and carbohydrate binding receptors are also thought to be involved in the regulation of cell growth. For example, transformed and mitogen–stimulated lymphocytes show increased levels of proteins antigenically related to lectins (carbohydrate–binding proteins). It has been suggested that the blood group active carbohydrates of the epidermal growth factor (EGF) receptors of cells could themselves be receptors for other endogenous regulators, distinct from EGF, which modulate the cellular response to EGF.[30] Furthermore, cell surface heparan sulphate proteoglycans are thought to control the access to cell surfaces of regulatory molecules such as growth factors, hormones, and neurotransmitters, and may influence local cation concentrations.[31]

Sialic acids, found in many oligosaccharides, are known to mask specific antigenic sites in cell surfaces and free molecules. This occurs, for example, on the surfaces of trophoblast cells where a sialic acid–rich glycoprotein layer provides an immune barrier between mother and embryo. This supression of antigenicity by masking antigenic determinants has been suggested as a major function of sialic acids and, indeed, partial loss of sialic acids may be the cause of some immune diseases.[32] We will see later the importance of charged groups such as sialic acids in the spreading out of saccharide branches and so contributing to this masking ability.

Growing evidence indicates an important role for the oligosaccharides of human chorionic gonadotropin (hCG). This is a dimeric glycoprotein hormone which exerts its effects on the cells of reproductive tissue by activating an adenyl cyclase system. Removal of the hCG oligosaccharides has no effect on its binding to specific cell surface receptors, but abolishes its ability to stimulate the production of cyclic AMP. Studies involving inhibition of cAMP production in rat corpora luteal tissue by hCG–derived glycopeptides indicate that activation of the adenyl cyclase system requires formation of a complex between the hCG receptor and a cell surface lectin. In fact, the hCG molecule provides the cross link, its oligosaccharide being bound by the lectin with the protein part binding to the hCG receptor.[9,33]

Cell–Substratum Interactions   Carbohydrates are known to participate in interactions between cells and extracellular matrices (substrata). Such interactions affect embryonic development, growth regulation, cell morphology, and maintenance of normal tisssue function.[24]

Proteoglycans (proteins with attached sulphated glycosaminoglycans) are especially important in this regard.[24,27] For example, cell surface heparan sulphate proteoglycans are thought to mediate the attachment and spreading of both stationary and moving cells. Cell surface proteoglycans are also thought to be the 'adheron' receptors responsible for the attachment of retina cells to substrates (adherons are supramolecular complexes comprised of the various proteins and proteoglycans). Growing evidence suggests that proteoglycans are also involved in controlling cell growth, basement membrane permeability, and the morphogenesis of nerve cells.[27]

Glycosaminoglycan binding is also found in fibronectin, a cell surface, extracellular matrix, and plasma glycoprotein which mediates cell attachment and spreading on collagen, fibrin, and artificial tissue culture substrates.

Cell–Cell Interactions   It is becoming increasingly clear that normality of cell surface glycoconjugates, as defined by the spatio–temporal distributions of specific sugars, is essential for normal cell behaviour. Oligosaccharides in secreted and cell surface glycoproteins and glycolipids show marked and reproducible changes during differentiation and development.[30,34] For example, stage–specific patterns of glycoprotein expression have been found in mouse embryos, mouse lymphocyte Thy–1, and differentiating human erythroid cells. Developmentally regulated changes in lectins have also been shown in slime moulds and vertebrate cells. Furthermore, inhibitors of glycosylation block embryonic development and differentiation. For example, development of tunicamycin treated sea urchin embryos is arrested at the gastrulation stage, whilst glycosyltranferase deficient mutants with altered cell surface glycosylation patterns exhibit reduction in adhesiveness.[35]

Alteration in cell surface oligosaccharides accompanies virus–induced, chemically–induced and spontaneous transformation to the cancerous state.[30,36] In BHK cells, changes in N–linked oligosaccharides of membrane glycoproteins accompanying polyoma transformation correlate with increases in a specific β–N–acetylglucosaminyl transferase. The fundamental relationship between altered cell surface carbohydrates and cancer implies an important role for oligosaccharides in immune surveillance.[28]

Surface membrane glycoconjugates have been suggested as decoders of extra–cellular information such that the response of a cell to surface interactions with substrata (e.g. fibronectins or collagens) or to other cell surfaces depends on their organization. The ways in which these messages could be transmitted into the cell will be discussed in detail later. There are several lines of evidence for the role of cell surface glycoprotein oligosaccharides as recognition molecules in cell adhesion.[37] For example, inhibition of protein glycosylation prevents compaction and trophoblast adhesion in mouse embryos. Cell adhesion glycoproteins have been found in developing nerve tissue, liver cells, cultured epithelial cells, fibroblasts, and embryonal carcinoma cells. These cell adhesion molecules (CAMs) appear in reproducible spatio–temporal order during cell development, undergoing changes in expression, prevalence and localization. Furthermore, they are thought to regulate various primary cell development processes. It has been suggested that polysialic acid found on glycoproteins of the developing brain may be involved in neural cell adhesion.[25,38]

Involvement of cell surface oligosaccharides has been shown in species–specific aggregation in several biological systems including sponges and slime moulds, fertilization in marine algae, invertebrates, the mating reaction of compatible yeasts, genesis and morphology in sea urchin and mouse embryos, the interaction of malaria causing parasites and host red blood cells, and the recognition of pathogens and symbionts. Surface carbohydrates are also involved in the binding to cells of infective agents, for example bacteria and viruses. Targeting of a parasite to host cells has been shown in *Giardia lamblia* in which a specific surface lectin, activated by secretions from the host's duodenum (the site of parasitic invasion) allows binding to the appropriate host cells.

Slime moulds have been particularly well studied as models of cellular differentiation. Cell surface carbohydrates and complementary receptors in these organisms have been implicated in cell–cell and cell–substratum adhesion, intercellular communication, and phagocytosis, and could act as membrane signals for differentiation.[39]

Growing evidence suggests the importance of oligosaccharide recognition in the immune system.[28] Carbohydrates are recognized by antibodies, complement components, cell surface lectins and enzymes, and carbohydrate–mediated uptake systems. Both humoral and cellular immune responses are regulated by a complex network of cellular interactions mediated by cell surface glycoproteins encoded by genes within the major histocompatibility complex (MHC). For example, asparagine–linked oligosaccharides have been shown to be important in the mixed lymphocyte response which is a model for cell–cell interactions for regulation of various immune responses. The 'homing' of recirculating lymphocytes to specific sites around the body is thought to involve recognition by homing receptors of specific cell surface oligosaccharides. It is also thought that regulation of cell surface glycosyl determinants at different stages of lymphocyte differentiation, together with appropriate distributions of homing receptors may be responsible for the placement of lymphocyte sub–populations within particular regions of lymphoid organs. Furthermore, functional subsets of lymphocytes can be readily separated on the basis of their specific reactivity to plant lectins. Recently, it has been shown that the N–linked oligosaccharides of human IgG change in patients with the autoimmune disease rheumatoid arthritis. The altered glycosylation pattern results from a 'population shift' to molecules with a reduced level of outer–arm glycosylation.[40]

Intracellular Routing of Proteins   Correct intra–cellular routing of proteins (for example those destined for secretion or incorporation into vesicles or membranes) is essential for maintenance of normal cell functioning. Recognition of proteins by the routing machinery probably involves many determinants (e.g. three–dimensional structure, specific protein surface 'patches', etc.). There is evidence that these include oligosaccharides. For example, the targeting of lysosomal enzymes from their site of synthesis in the rough endoplasmic reticulum to their final destination in lysosomes involves a series of specific recognition events. Phosphomannosyl residues on lysosomal enzymes have been shown as essential components of the recognition marker required for binding to intracellular mannose–6–phosphate receptors and translocation to lysosomes. Also, maturation of certain secretory glycoproteins from the rough endoplasmic reticulum to the Golgi complex requires appropriate processing of N–linked oligosaccharides. [41]

Species– and Organ–Specific Glycosylation   Species– and organ–specific glycosylation has been shown in many glycoproteins.[4,9] These include rat and human α_1–acid glycoproteins, bovine and human prothrombin, and bovine and human C1q. Extensive studies reveal that the glycosylation patterns of γ–glutamyltranspeptidase shows marked organ specificity.[42] Interestingly, the neutral (asialo) cores of N–linked oligosaccharides of immunoglobulin G (IgG) from human, sheep, rabbit, horse, and cattle sera all show different proportions of the same set of complex–type bi–antennary structures.[43]

Microheterogeneity and Site–Specific Glycosylation   A striking feature of glycoprotein oligosaccharides is the large array of different oligosaccharide structures often found even in material from homogeneous cell populations. This phenomenon is usually termed microheterogeneity.[4,9] Examples include hen ovalbumin which has at least sixteen different oligosaccharide structures found at a single glycosylation site and myeloma IgM heavy chains, human ceruloplasmin, rat kidney γ–glutamyltranspeptidase and avidin. Furthermore, this microheterogeneity has been shown, for some systems, to be stable. For example, sialylation and branching patterns of N–linked oligosaccharides of mouse major histocompatibility antigens are site specific and reproducible.[44]

Recently, it has become clear that by controlling glycosylation different tissues can create unique sets of glycoforms (i.e. glycoproteins with identical polypeptides but with oligosaccharides that different in either sequence or position on the protein backbone).[45] For example, a study of the cell surface glycoprotein Thy–1 isolated from rat brain and thymus has shown that for both tissues the molecules bear three glycosylation sites (Asn–23, 74, and 98) but there is tissue–specificity of glycosylation upon which is superimposed a significant degree of site specificity. In fact, on the basis of the site distribution of oligosaccharides, one finds that there are no Thy–1 molecules in common between the two tissues, despite the amino acid sequences being identical. It is quite possible that the discrete subsets of glycoproteins corresponding to these glycoforms have different physical and biochemical properties leading to functional diversity.

The short review of saccharide functions indicates the extensive evidence of the importance of these molecules in biological systems. The question arises as to those molecular properties of saccharides which have led to their use, rather than other biopolymers, in these functions. To cover the vast variety of poly– and oligosaccharides their classification shall here be followed as related to these functions, especially in association with membranes. First, we examine the value of long chain linear polysaccharides with simple repeating units in a given mobile array, say 1→6 condensation, assuming little interaction between units. We shall then ask what effect is there of the introduction of different side chains which allow greater or lesser chain/chain interaction whether this is internal to a chain or between chains. These polymers will be shown to generate bulk physical properties of great functional value but not related to chemical functions such as molecular recognition.[46]

The next complication is to introduce branching (effectively terminal 'cross links') and then, finally, to add specific side–chain functions to the branches. Here we approach the more defined three–dimensional forms of the shorter chain oligosaccharides. At each stage of discussion we look at structural restrictions, internal mobility, and chemical functionalities. At first, the classes will be treated in a general way in order to establish principles. This is followed by specific examples.

9   TYPE I LONG LINEAR POLYSACCHARIDES

Random Structures

In descriptions of bio–polymers we are now used to the notion of structure as in the chemical crystalline state e.g. the structure of proteins. This is valid for certain types of polysaccharide, but another useful starting point for a quite different type of polysaccharide is the perfect gas or dilute solution model of random motions. The appropriate structure/dynamic approach is given in texts on synthetic polymers such as polystyrene.[47–49] One can start from concepts such as the osmotic pressure, π, where one writes, without reference to molecular structure,

$$\pi = RTc \tag{1}$$

where c is the concentration of solute in moles per litre, R is the gas constant, and T is temperature. Next, we take into account either repulsive or attractive interactions with solvent and between molecules in the second and further virial expansion terms, A.

$$\pi = RT(c + Ac^2 \dots)$$ (2)

The very fact that the polymers are made of large monomeric units causes there to be an excluded zone (often called the excluded volume) and the factor A also includes this effect. Looking at this more carefully we see that for a polymer molecule the 'concentration' depends on the ratio of molecular volumes of individual polymers to the volume of a molecule of solvent and that this ratio is not fixed for polymer molecules since there are internal variations in volume with concentration. The amount of free solvent varies depending on the degree of swelling of the polymers. These two are interactive terms so that there are additional terms in the virial expansion due to the internal constructions in single polymer molecules. The restriction on space and then the solvation of the polymer is limited by its free volume at a fixed concentration.

A parallel starting point for the dynamics of molecules in solution is the viscosity of a polymer solution. Ideally,

$$\eta = \eta_s (1 + c\eta)$$ (3)

where η_s is the solvent viscosity. Once again, it is necessary to take into account the internal and external interaction between polymer molecules and solvent and, assuming them to be in equilibrium, we expand the viscosity equation

$$\eta = \eta_s (1 + c\eta + Bc^2\eta^2 + \dots)$$ (4)

where the higher order terms, B, etc., will be related to internal and external interaction dynamics with solvent. Unfortunately, there is insufficient space in this article to deal with the properties of the mutual flow of polymers and solvent which is so very important in biological movements, e.g. on the outside of aquatic animals or in the alimentary canals of all animals.

In dealing with these bulk properties one should stress that polymer functions for polysaccharides are very different from those most usually considered for proteins. Open polysaccharides sense the solvent restrictions around them and are freely adjustable molecules. This means that they can sense temperature for example (equation 1). The value of this is shown below.

Now, when the solutions become concentrated the polymers lose freedom of movement except locally due to entanglements from which they cannot escape. The properties of entangled non—interacting polysaccharides, for example in

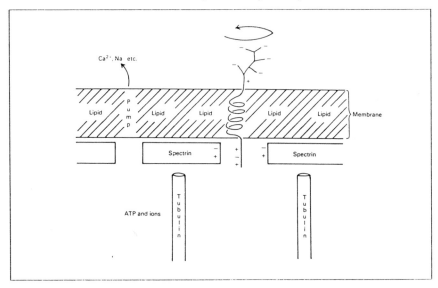

Figure 3 : Receptor Signalling   A schematic representation of a possible connection between a random polysaccharide on a protein and the internal network of cellular proteins.

mucous secretions, needs a theory such as the tube theory of Doi and Edwards.[49] These entangled polymers which are held to a floating frame (e.g. a cell membrane) will be restricted in volume by any surface they meet. In other words, work is done when the polysaccharide is constrained to a smaller volume than the random walk volume.[49] In effect, a random polysaccharide measures available volume (equation 1). If the polysaccharide is bound to a protein then it will carry the protein along as it moves into free solvent space away from such a constriction. Effectively, it will bounce away from a constraint. If the polysaccharide is bound to a deformable object then restriction of the volume of the polysaccharide will act as a deforming force on the object, say a membrane, so as to gain volume for the polysaccharide. If the polysaccharide is bound to an adjustable unit embedded in a rigid matrix (membrane) then the constraint on the volume of the polysaccharide will drive the adjustable unit through the membrane. This is a plausible mode for receptor signalling action as we shall see (Figure 3). Here, the random walk polysaccharide can act as an entropy device for measuring constraints on the free volume outside a cell or for measuring temperature (see equations 1 and 2). When considering polysaccharides one should not forget the deep penetration of water into the polymer so that the two form a system which can be quite rigid as in the binding of glycopeptides to ice, or very loose as in the above.[17]

Packing and Cell Shape

Another way in which expanded regions of polysaccharides attached to proteins in membranes may be responsible for, or related to, volume restrictions is through spatial packing which relates to shapes of cells. For a spherical cell the surface tension of the membrane is uniform so that packing of molecules in the membrane is only radially different i.e. from inside to outside. In the case of a membrane with regions of different curvature then local surface tension is different if the composition of the membrane is uniform. As a consequence of this, components of the membrane will tend to sort themselves so as to reduce the

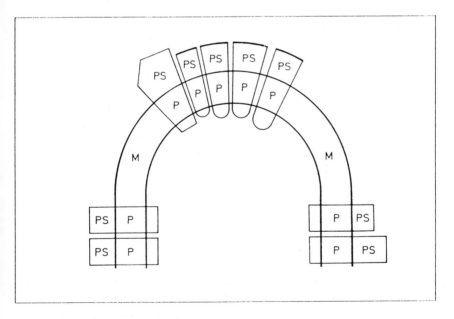

Figure 4 : Membrane Tension   The polysaccharides, PS, are shown defining the curvature of a cell surface by virtue of the way in which their size and shape dominate packing.

differences. This differential packing of molecules in curved and straight zones is shown in Figure 4. Conversely, the presence of differently shaped molecules can *cause* the curvature. Packing and shape now involve an interaction between external forces and the tensions of filamentous structures internal to the cell in contact with the surface. The function of different polysaccharides of membrane proteins could then be to assist localization of protein activities in zones along the inside as well as the outside of the membrane. Finally, polysaccharides can be used to provide a loose platform for many kinds of biomineralisation.

10 TYPE II LINEAR POLYSACCHARIDES WITH CHAIN–CHAIN INTERACTION

In this section we shall consider secondary and tertiary structures of polysaccharides. A limitation away from a random coil can be based on repulsive or attractive energies as well as simple space filling considerations. Examples include :

(a)   Anionic linear polysaccharides (repulsions)

(b)   Linear polysaccharides interacting through H–bonds (attractions)

We shall only look at hydrophobic forces briefly as these do not appear to be of such dominant importance in polysaccharides as compared with proteins and polynucleotides. Each new force is an extra term in the virial expansion of equations (1) and (2) but it also introduces some structural chemical specificity, much of which now depends on selected side chains.

Anionic Linear Polymers [50]

A repeated anionic unit in a linear polymer will tend to stretch out the molecule (cf. DNA). Such a structure will resist constraints on its volume as for a random polymer but now proportional also to the increase in electrostatic repulsion. This repulsion also pushes the polysaccharide away from any negatively charged surface. It will sense long–range electrostatic fields as well as short–range volume constraints. However, as a rod, it can sense direction too. If a more general volume detector is required it must be made from several mutually repelling rods on a protein or membrane frame. The rod will also sense ion concentrations in the external solution due to two interactions which affect its volume:

(a)   the general ion effects (Debye–Huckel–Onsager)

(b)   the specific binding (cation) effects.

General theories of ion interaction with polyelectrolytes have been developed (see reference 51). These interactions can cause collapse of a widely dispersed anionic polymer molecule over a very small concentration range of cation to a very small volume. Furthermore, like all other attractive forces they give rise to equations of state parallel to those of condenseable non–ideal gases. This is il-lustrated in Figures 5 and 6. So far, these have been considered as free units. Now one must ask what value they could have when bound to a surface.

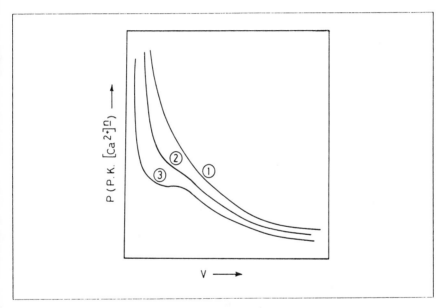

Figure 5 : Phase Diagram A typical PV phase diagram substituting calcium ion concentrations for P and the molar volume of a polysaccharide for V. The ideal case is the top curve. The bottom curve illustrates phase separation.

Polymers such as these will not be very useful as sensors on the surface of single cells if the cell is a sphere and membrane–bound molecules can move over the surface such that the cell remains isotropic. In such a case, escape from repulsion by a surface of the same charge could be managed by diffusion of the anionic

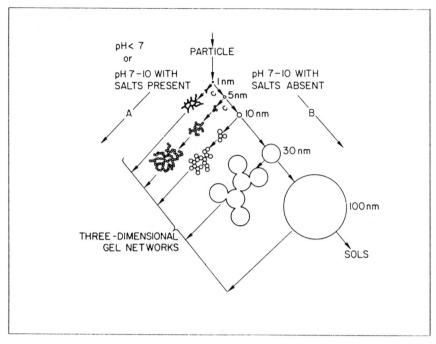

Figure 6 : Gels and Sols   A typical structural diagram of gel/sol changes with ionic concentrations. (After reference 46).

polymer along the membrane. A re–arrangement of glycoproteins would occur but the cell itself would not be repulsed. A more rigid anisotropic cell is more interesting. For example, the red blood cell has a discoidal shape that could not arise if the membrane was under isotropic constraints. Furthermore, its volume is not restricted since such constraints would generate a sphere.[52] In fact, the shape is maintained by polymeric protein networks under the membrane (see Figure 7) and energy is expended to keep the cell stable. The membrane shape requires differential radii of curvature at different points. This, in turn, means that there will be differential forces along the membrane and these forces are directly related to curvature. However, since the membrane is not an homogeneous chemical system the curvature stresses in it can be partly relieved by lateral molecular concentration gradients such that some molecules (e.g. glycoproteins) are more concentrated in some regions (of given curvature) while others are more concentrated elsewhere (Figure 4). In essence, a shaped membrane can relieve its tension by re–arrangement of molecules. When such a cell strikes a constraint the interaction cannot easily be relieved by protein diffusion round the cell surface. Instead, a pressure is communicated *into* the cell, <u>i.e.</u> signalling will occur (see Figure 8). A message enters the cell when a mechanical pressure or an electrical repulsion is exerted first on the extended (charged) polysaccharides and then, say, *via* a membrane protein helix (adjustable in an up or down sense) to the intracellular protein trigger.[53] The cell will now change shape, these processes perhaps being assisted by metabolic events communicated to stress fibre systems including the opening and closing of channels in the membranes (see Figure 8). The cell then moves its membrane away from the constraint. By this mechanism, a red cell under pressure from the flow of blood for example could squeeze into a capillary smaller in diameter than the initial size of the cell itself. As stated above, this type of sensor can be directional, especially if the sensor is rod–like and carries charge.

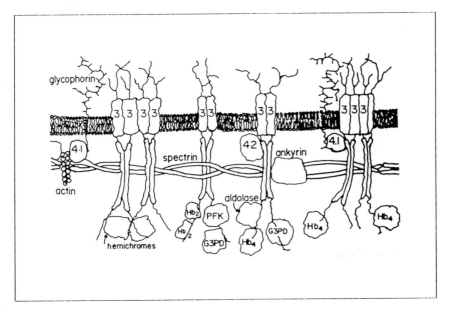

Figure 7 : Membrane Protein Networks   A more realistic description of the erythrocyte, see fig. 3. (After reference 52)

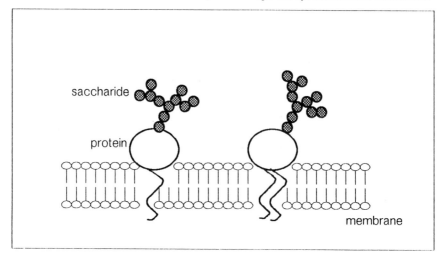

saccharide

protein

membrane

Figure 8 : Push–Pull Mechanisms   The structure on the left represents a glycoprotein connected to a single helix device (push–pull signal) while on the right the connection is to a multi–helix channel signal. Compare these to the membrane–bound glycoproteins in the immunoglobulin 'superfamily' shown in reference 55.

The interaction with a surface may also be attractive due to the attachment between negative surfaces and negative polysaccharides through the mediation of calcium ions. Extension (but not retraction) of membranes into pseudopods is aided by such interaction. Cell–cell interactions made in this way generate communication networks between cells.[46,54] These are positive communications similar in essence to those generated by repulsion as described above. In part, this binding is related to the sol–gel changes produced by ions (see Figure 6).

In practice, one observes many carboxylated and sulphated polysaccharides with a wide variety of composition giving the same function (see chondroitin sulphates). Some of these functions are those typical of non–biological polymers e.g. adhesives, flow assistant fluids, etc., but here one should stress the volume of the polymers as sensors and note that these are properties not often associated with proteins.

Long Chain Polysaccharides Interacting via H–Bonds

The nature of attractive, secondary, and tertiary constraints from the coming together of (neutral) polysaccharides must now be examined. Since polysaccharide/polysaccharide interaction based on charge neutralisation is rare due to the infrequency of positively charged side-chains, there remain two obvious cases:

(1)   –OH H–bonding to other OH groups and

(2)   the –NH of side chain amide to carbonyl bonding.

The neutral polysaccharides can form helical copolymers or β–sheets respectively from 1 and 2. Examples will be given later, but immediately we notice that water is excluded from these polymers. These structures are really combinations of extended strings like DNA and RNA and unlike proteins since they have no substitutions on the side chains to make them fold into particular shapes with strict locations for functional groups. Finally, they are usually made with a very small variety of types of saccharide and the limited number of folded structures is striking.[1]

These polysaccharides are dense, strong structures resembling crystalline solids and are open to X–ray diffraction studies, with many such structures known.[1,17] Structural fidelity is important for these molecules, but is not easily achieved by enzymatic protein synthesis without a template unless the number of different monomers is strictly limited. Notice the use of A_n or $(AB)_n$ repeats in these systems and contrast this with the infinite variety of sequences of DNA or the great variety of sequences which give rise for example to the haemoglobin fold. Also, note the variety of sequences in the shorter glycoprotein oligosaccharides.

11 ORDER PARAMETERS

We would like to describe different polysaccharides by their dynamics as well as by the conformations which they can adopt (see for example references 48,56). One way of doing this is to consider the sum of the probabilities of different conformations, Z, open to one molecule of a polymer multiplied by the degeneracy of that conformation and weighted by the energy ΔE_i required to reach the conformational state from some lowest ground state (Figure 9) :

$$P = \Sigma_i P_i = \sum_1^Z n_i \, e^{-\Delta E_i /RT}$$

The probability of being in a given state, i, is then the P_i for this state over this sum. For a folded rigid polymer there is one (ground) state which has a probability P_i of one. We now plot on a logarithmic scale a diagram of systems of different numbers of states, ΣP_i, indicating the degree of order $1/\Sigma P_i$ for polymers of x units. In order to see the importance of this diagram we shall at first assume that $-\Delta E_i /RT$ is very small so that the exponential term in the above equation tends to unity and relate P to the number of states of equal energy, <u>i.e.</u> the number of configurations. It is well known that the ability to act as an energy absorbing system is directly related to total configurational entropy

$$R.\ln(\Sigma_i P_i) = \Delta S$$

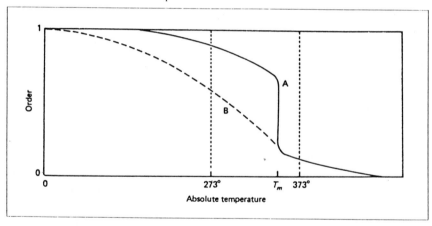

Figure 9 : Order Parameters   The variation of order in a virtual solid <u>e.g.</u> a protein (curve A) showing the typical melting behaviour and the order changes and in B a more glass–like disordered system <u>e.g.</u> some polysaccharides.

Those polymers which could act as non–specific three–dimensional devices for the measurement of fields (volume, chemical, or electrostatic constraints) or temperature, <u>i.e.</u> those described immediately above, are at their best if ΔS is large but then they shall be of large Z and are structureless (compare with a perfect gas). Most often in biology some structure is required. The more that is required in the way of well defined structure and its recognition the less that a large variety of states of equal energy can be afforded, so the polymer should have small Z. Recognition in a reasonable period of time demands low ΔS but some mobility is required in order to bind quickly. The compromise within activity depends on how selective the recognition must be to perform the function within a given time span. The more exacting the fit the slower the fitting must become. Since biological systems must respond in a relative short time they are required to have compromise polymers which may occupy several states in order to react quickly. Here, saccharides are ideal but we can see that selection with regard to function is based on chain length and relative rigidity and flexibility and not on folding.

One can analyse the probability that a polymer which has monomers with rotational degrees of freedom around n bonds will be in a given state as follows. Suppose that at each of the n bonds there are 3–rotameric forms of equal energy and that all other states are of too high an energy to be occupied. We then find that the probability of the single state of interest is $(1/3)^n$. The value of this term is an entropy off–setting the probability of recognition of a particular conformation by a rigid receptor. A polysaccharide of x units has the ether linkage with two bonds which are basically rotationally free when the probability of the given state is $(1/9)^x$. We must then include the side–chains, one of which may be $-CH_2-X$ while the others are rigid, when we have a probability of the given state of $(1/9)^x(1/3)^n$. If we suppose that recognition is based on minimally three units and maximally six, each with one single side–chain, we find the entropy term against recognition of one particular state $R.\ln(\Sigma P_i)$ is for three units 16 cal/mol/K and for 6 units is 32 cal/mol/K. This begins to be a considerable randomness factor against recognition. $T\Delta S$ is approximately 5 kcal/mol or 10 kcal/mol respectively at 300 K when the adverse entropic factor for binding is about 10^4 or 10^8. Clearly, if a large number of saccharide units are involved in recognition then constraints on mobility are necessary in a recognition system. The thermodynamics of sugar–receptor binding are discussed further in a later section.

Another possible use of a polysaccharide is to provide a solubilising protection for a protein or a cell so that it is *not* recognised. Examples are some polysaccharides of plant peroxidases and red blood cells and we have already encountered the masking ability of sialic acid rich branched oligosaccharides. Here the use of long, highly random polymers would be advantageous as they are not easily recognisable.

Finally, a polysaccharide which is a dynamic random coil occupies a considerable volume. It is then possible to use it as a sensor of space and temperature fields around a cell. Moreover, these more random polymers can adjust extremely rapidly because they have many closely related states. They are excellent membrane protectors and sensors. Returning to the above equation we see that temperature has an important influence but we can not expand on this feature here.

Before turning to selective recognition another point is clear. If relatively small branched chains are made and particular branches are used then the surface area is increased but there is an inhibition of the space available, i.e. ΣP_i is lowered. We now look at branched chain saccharides since they are the principle recognition device in glycobiology systems.

12 BRANCHED CHAINS

Branched chains give the following advantages assuming a single fixture point at a base:

i   Greater potential for diversity close to the base i.e. development of recognition features at low degrees of polymerisation and close to the holding unit (protein or membrane).

ii   Greater coverage of the surface close to the base i.e. the membrane or protein at the base so providing good protection (against proteolysis for example).

iii   More direct interaction with the base i.e. the cell membrane supporting the structure. In effect, relaying information from the extremes of the umbrella to the handle is made direct. A long, possibly random, stalk to a recognition system could dissipate the energy of recognition without propagation down the stalk to the handle. The stalk (core) should be relatively rigid, e.g. $1 \rightarrow 4$ connectivity, not $1 \rightarrow 6$ (see Figure 2).

The branched structure has the following disadvantages:

i   It takes up space relative to the base.

ii   It does not reach as far from the base per monomer unit as a linear structure.

iii   It has a lower probability per unit of occupying more than one conformation since the branching is a steric or volume restriction (see below).

The number of states occupied by branched chains will be further reduced by
introducing charged substituents. Usually such charges are found at the ends of
the branches (for example as sialic residues) so that the branches tend to be
spread apart. Note that branching chains tend to give convex surfaces suitable for
interaction with concave receptors and that folded proteins easily generate concave surfaces (grooves) but saccharides do not. Note also that, in contrast to long
linear polysaccharides, branched chains do not make effective entropic sensors.

Fidelity of structure now becomes of extreme importance. It matters greatly in
recognition if even one −OH is misplaced or substituted. Given the method of
synthesis (i.e. without a template) we can expect to get strict fidelity by combinations of the following :

i   use of simple sequences of repeated units, e.g. $(mannose)_n$,

ii   use of a limited number of units to reduce the possibility of error and
keep the entropy low, and

iii   use of coordinated, strictly regulated biosynthetic enzyme 'factories'.

13 THERMODYNAMICS OF SACCHARIDE−RECEPTOR BINDING

Three distinct interactions can be expected to contribute to the thermodynamics of
association of neutral oligosaccharides with receptors (see reference 57 et cit).
These are hydrophobic, H−bonding, and van der Waals forces. There is evidence
that aromatic amino acid side−groups are important in hydrophobic interactions
between protein receptors and D isomer sugars. This arises in part from the
monosaccharide units adopting a $^4C_1$ D conformation in which the planar
hydrophobic face can interact with the planar aromatic rings. Such interactions
are seen in sugar binding immunoglobulins and the binding of lysozyme to its
sugar substrate. Regarding hydrogen bonding, sugars in aqueous solution are
extensively H−bonded with water. Transference of these H−bonds with solvent to
a space within a receptor site which is itself hydrated should not contribute greatly
to the thermodynamics of association. However, an important contribution could
occur if the H−bonds of the complex were formed in an environment of low

dielectric constant, for example in a combining site with a predominance of non-polar amino acids. The contribution of van der Waals forces to sugar binding is difficult to estimate. No doubt, the loss of conformational entropy associated with the binding of, for example, a tetrasaccharide would make the ΔS of association unfavourable. However, this would be offset by the structured water of hydration previously associated with both protein and ligand which would be returned to bulk solvent.

A basic feature of sugar binding proteins is the concept of distinct subsites, the shape and topology of which will determine the specificity for a particular sequence of sugars.[57] For example, an individual subsite which may be aromatic might not distinguish between sugars and other hydrophobic (e.g. aromatic) ligands which have the correct shape. However, the specificity of the sugar binding protein would arise from the presence of several sub–sites. This is shown by the increased strength of binding of tetra– and tri–saccharides compared with that of monosaccharides to sugar binding myeloma proteins and to lysozyme where, for example, tri–saccharides bind three orders of magnitude more strongly than the monosaccharides. These sub–sites therefore favour carbohydrates which are often polymeric and discriminate against smaller hydrophobic ligands which bind to individual sub–sites, but with lower affinity.

Lectins are an interesting class of sugar binding protein in which each molecule contains at least two binding sites for specific ligand domains. One can recognise two classes of these proteins based on the nature of interaction with the saccharide ligand. [58] Class I (simple binding mode) lectins possess subsites for specific monosaccharides of the ligand while Class II (complex binding mode) lectins have subsites for specific carbohydrate sequences, none of the individual monosaccharides in the sequence playing a predominant role in the binding process.

14 SELECTED EXAMPLES OF THE DIFFERENT TYPES OF SACCHARIDES

Neutral Linear Chains

There are many neutral linear polysaccharides. The $\alpha 1 \rightarrow 4$ linkage in glycogen (where only about one in ten residue branches via an $\alpha 1 \rightarrow 6$ linkage) allows the extended structure to be packaged in granules as a relatively open mobile ball of neutral storage forms of sugars in plants and animals. These structures are highly hydrated. Note the form of the polymer is $(A)_n$.

A very different package is made from the $\beta 1 \rightarrow 4$ units of cellulose, chitins, and bacterial cell walls which have no branches. They all form very strong structures resembling, in the case of chitin, the β-sheets of proteins in which mutual H-bonds give a compact structure from which most of the water is removed. Many of the other structures are parallel runs of helices (cf. collagens) and these units can move as a single piece only.[1] Thus they can be used in a variety of ways structurally. Note they are $(A)_n$ or $(AB)_n$ polymers. Once again, very few motifs can be generated unless we fold the helices and sheets together in different ways. As we have seen, this does not occur with polysaccharides.

Charged Linear Polysaccharides

The example here is the sulphated glycosaminoglycuronans.[50] The chain is condensed mainly in $1 \rightarrow 4$ linkages with some $1 \rightarrow 3$ linkages and there is no branching. The chains are often extensive ($n > 20$), sometimes as long as 100 units. Given the condensation pattern they are likely to be stretched linearly away from the protein to which they are bound. The stretched out conformation is very frequently re-inforced by extra anionic charges, often two per residue and usually $-OSO_3^-$ groups but now and then there are $-CO_2^-$ groups where the residue is iduronic acid. Although the saccharide monomers are aminoglycurons they carry no positive charge since the amino group is invariably acetylated. The crowding of negative charges along the polysaccharides parallels that in DNA and RNA which form helical extended structures and which have considerable local charge and are specific counter ion dependent in their conformational properties. While the $1 \rightarrow 3$ and $1 \rightarrow 4$ disaccharide links are relatively rigid compared to the phosphate linkages and the ribose rings of DNA and RNA they have some degree of conformational flexibility under different counter-charge concentration conditions. We have mentioned in the section on anionic linear polymers that such flexible polysaccharides can collapse under certain conditions (see Figure 6).

What is the functional significance of these polymers? It seems they are not principally points of recognition but, rather, are designed to occupy space in a particular way. In large part they will be surrounded by water. The $-OSO_3^-$ group is very poor at forming H-bonds and it does not bind metal ions, even Ca^{2+} and Mg^{2+}, effectively at the levels of these ions in biological solution.

These polymers seem designed to form an open mesh (sol) around cells and connective protein fibres so as to allow :

i   easy diffusion of small or medium sized molecules to and from cell surfaces,

ii   very restricted access of such cells to one another, and

iii   useful mechanical properties.

In addition, it is clear that the polymers will repel one another and will occupy large volume elements on the outside of cells relative to the inside. Such a structure causes membrane curvature (Figure 4) and it appears that they are intimately related to cell volume and shape as well as to dynamic (flow) functions. Alone, their function cannot be connective, but in combination with connective tissue molecules such as collagen or surface adhesive molecules such as CAMs they will maintain open aqueous zones around connected cells. Perhaps this is why they are so essential in multi-cellular organisms.

Branched Chains

Long Chains   We take glycophorin as the first branched chain example. NMR data shows that the long polysaccharide chains are highly mobile relative to the protein core and that they form a branched umbrella structure. The manner of their linkage to the underlying contractile system of the cell is relatively well defined (Figure 7). It would appear that they are ideally constructed as probes of space around the cell. Not only will they detect volume interference by virtue of their long branched chains but they will detect surfaces through the charged terminal neuraminic acid units. A number of glycoproteins and glycolipids may be used in this simple general physical detector role rather than for the specific chemistry to which we now turn. Very strict structural fidelity may be of little consequence to these devices except at the terminii.

Short Chains : Oligosaccharides   The second example is the branching oligomannose oligosaccharides found on many glycoproteins (Figure 10). Starting from the protein, we observe the following:

a   On the protein surface mobility is usual around the β–CH_2– group of the asparagine, not at the amide

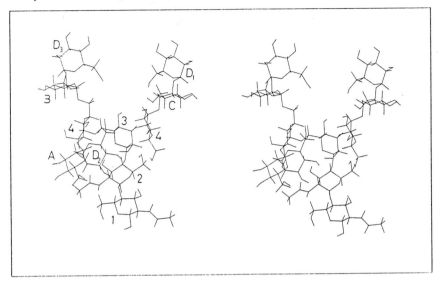

Figure 10 : Stereo Picture of an Oligomannose oligosaccharide
The conformation of the structure was obtained from a combination of NMR experiments, molecular orbital calculations and molecular dynamics simulation (see reference 14). The primary structure and numbering scheme are as follows:

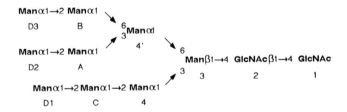

b   1→3 and 1→4 links are relatively constrained

c   1→6 links are relatively free

d   1→2 links are constrained and also bend the structure.

e   The chains are relatively short.

We notice immediately that the branching pattern at the β—mannose or the α—mannose of a typical oligosaccharide (see for example Figure 2) gives a rigid 1→3 arm and a mobile 1→6 arm. We also notice that a series of 1→2 substitutions effectively require the saccharide to turn back on itself. 1→6, 1→3, or 1→4 substitutions can give elongated chains. Self—interference volumes then restrict the number of possible conformations around the ether links especially for 1→2 links. Looked at in this way, we see that the first three linear residues (comprising the mannosyl chitobiosyl core) and the first two residues after a branch stretch the structure away from the protein but the next units turn to cover a surface at some distance from the protein anchor point in the membrane. This is an excellent structure for expressing specificity while being built on a rod. The stereo picture of the structure obtained by a mixture of NMR and energy calculation studies is shown in Fig. 10. Notice that if all three arms of the saccharides are to be used then taking the initial stem from the protein as a fourth arm an overall shape deviating from a tetrahedron toward a narrower cone angle is of greater value as an exposed recognition site than a system of a wider angle. However, a wider angle or even an inverted angle will cover the support protein more successfully. Note that very few monomeric units are required to express a selective surface without folding and contrast this structure with the fold of proteins. Looking back to section 8 we can see how these properties contribute to the many roles of oligosaccharides in biological systems.

An interesting type of branching is seen in plant oligosaccharides where a fucosyl residue branching close to the base probably constrains movement of the core structure :

$$\begin{array}{l} \text{Man}\alpha1 \\ \qquad\qquad\searrow\ 6 \\ \text{Man}\alpha1\rightarrow3 \quad \text{Man}\beta1\rightarrow4\text{GlcNAc}\ \beta1\rightarrow4\text{GlcNAc} \\ \qquad\nearrow\ 2 \qquad\qquad\qquad\qquad\quad\nearrow\ 3 \\ \text{Xyl}\ \beta1 \qquad\qquad\qquad\qquad\text{Fuc}\alpha1 \end{array}$$

Another interesting case is the penta—antennary oligosaccharide from hen ovomucoid where multiplicity of branching shows clearly the ability of a branched saccharide chain to express a recognition surface :

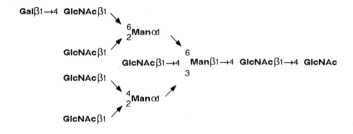

15 SPECIAL SACCHARIDES

Apart from the common monosaccharides such as <u>N</u>−acetyl−glucosamine, galactose, and mannose, there are a host of unusual monosaccharides which are specific to particular organisms. For example, the shell of the coccolith contains very complex saccharides. It is not just the branching which is important but the nature of the individual units. The problem of the selectivity involved in their uses arises frequently in the description of surface/surface contacts, <u>e.g.</u> a special saccharide is involved in the invasion of *E.coli* by phage.[59] No doubt, special saccharides are involved in other types of invasions, in cancer cell surfaces, an so on.

16 CONCLUSION ABOUT MOBILITY AND STRUCTURE

It follows from the analysis of mobility within structure that one should expect different types of polysaccharide to have different functions. One idea is that a polymer which is attached to a membrane and which is large, very open, and mobile can exert an increasing force as a sensor while itself decreasing its volume (an entropy sensor or a detector of space, temperature, and chemical and electric field restrictions). Clearly, it could also be used as a chemical ion gradient sensor with a graded response if the chemical, <u>e.g.</u> calcium ions, interacted with the polysaccharide. This should be called a *bulk physical recognition device*. Polysaccharides can also be used as *physical flow* aids but we have not discussed such properties here.

At the other extreme, imagine a rigid polymer recognised by an unique rigid receptor. This is the dye in the mould (sometimes called lock and key) recognition device. Now for complex surfaces such as those presented by saccharides this mode of recognition is inherently slow since it requires an all or none recognition. A more appealing device is hand–in–glove recognition where the two surfaces have some relative mobility. This allows gradual marrying of the surfaces and so greatly enhances the on– and off–rates for the same binding constraint. Given the difficulty of forming a fold of a saccharide, the best device using it is a relatively short, branched chain.

Work on such oligosaccharides (summarised in references 12 to 14) has shown that they exist in solution with regions of restricted three–dimensional structure. Within these structures certain key residues modulate the overall conformation. For example, addition of a 'bisecting' GlcNAc to the complex biantennary oligosaccharide as obtained from human serotransferrin restricts the $\alpha1{\rightarrow}6$ antenna to be folded back towards the sugar's core; in the absence of this residue the arm can exist in two equilibrium positions, i.e folded forward or folded back.[10] Taken together, these two results show that the potential for joint rigidity/flexibility can exist in these molecules. The suggestion must be that the mobile arm (finger) helps to 'feel' the way into the receptor (glove). If the receptor is a protein then it is unlikely to have significant mobility of its backbone. It is then the ability of side–chain protein mobility together with backbone (and some side–chain) mobility of oligosaccharides which is the basis of the fast molecular recognition process. A simple case is the small mobility of the surface of lysozyme as shown by the flipping of Tyr and of Val, and the flapping of Trp as compared with the mobility of the hexasaccharide substrate composed of $1{\rightarrow}4$ linked GlcNAc residues. The fitting itself is based on molecular shape, bonding, and chemical H–bond matching. This should be called a *molecular chemical* recognition device. Notice again the degree to which the structure is decided by space filling rather than by positive interactions as in proteins.

Given that saccharides as a class of polymer can fill all the roles between the extremes of an *entropic physical recognition device* and a *molecular chemical recognition device* because all degrees of mobility and rigidity can be attained (but they cannot carry activated catalytic side chains such as imidazoles, thiols, and transition metals) it is highly appropriate that they dominate extra–cellular space and exert their influence by interaction through proteins in membranes to which they are attached but they are not involved in chemical transformations.

Let us now extend the description to the manner of transferring the sensor changes into messages within a cell. We have pointed to two means :

i   The first mechanism involves a push–pull on a simple (helical) protein strand which runs from the sensor through the membrane and then contacts the internal apparatus of a cell. In this case, no other message except one of conformational change is transmitted to the cell. Example could include glycophorin or the EGF receptor, both of which have a simple helix in the membrane.

ii   The second case involves a more complex unit composed of a series of helical protein strands connecting the polysaccharide through the membrane to the inside of the cell. In this device the sensor must impart a twisting as well as a lateral motion to open an ion channel. Here the transmitted message is diffusion of a chemical. A possible example is given in Figure 8.

In these cases, steps such as kinase activation in the membrane or the cytoplasm are regarded as secondary to the sensor activation.

Finally, we note that as recently as 1966 in a paper entitle 'On the theoretical role of glycoproteins', Eylar wrote

"It is a striking feature of glycoproteins in general that the carbohydrate appears to have a structural and not a functional significance". [60]

Nevertheless, he proposed that oligosaccharides act as chemical labels which, upon interaction with membrane receptors or carriers, promote the transport of newly synthesized glycoprotein into the extracellular environment. Twenty years later, a panel of scientists convened by the US Department of Energy concluded that :

"Understanding the roles played by complex carbohydrates is the next great frontier in the advancement of molecular biology". [61]

Increasingly we are seeing the great importance of these molecules in biology. In this article, we have only touched upon the very special nature of saccharide chains that makes them in very different ways extremely valuable parts of the apparatus for communicating between a cell and its outside environment or between cells.

Acknowledgements

The authors would like to thank Prof. Raymond A. Dwek and Drs. David Ashford, Chris Edge, Raj B. Parekh, and Thomas Rademacher of The Glycobiology Unit, Oxford University for their help and suggestions.

References

1 D.A. Rees 'Polysaccharide Shapes', Chapman and Hall, London, 1977.

2 M. Hook, L. Kjellen, S. Johansson, J. Robinson, **Ann. Rev. Biochem.**, 1984, **50**, 555

3 W. Pigman and D. Horton, 'The Carbohydrates. Chemistry and Biochemistry Vol IA', 1972, Academic Press

4 R.C. Hughes, 'Glycoproteins', 1983, (in 'Outline Studies in Biology' Series, Chapman Hall)

5 A. Kobata, in 'Biology of Carbohydrates Vol 2' (Eds. V. Ginsburg, P.W. Robbins), 1984, John Wiley and Sons, New York

6 P.V. Wagh and O.P Bahl, **Crit. Rev. Biochem.**, 1981, **10**, 307

7 J.E. Sadler in Biology of Carbohydrates Vol 2' (Eds. V. Ginsburg, P.W. Robbins), 1984, John Wiley and Sons, New York.

8 J.F. Kennedy, C.A. White, 'Bioactive Carbohydrates In Chemistry, Biochemistry and Biology', 1983, Ellis Horwood.

9 T.W. Rademacher, R.B. Parekh, and R.A. Dwek, **Ann. Rev. Biochem.**, 1988, **57**, 785

10 S.W. Homans, R.A. Dwek, J. Boyd, M. Mahmoudian, W.G. Richards, and T.W. Rademacher, **Biochemistry**, 1986, **25**, 6342

11 K. Nakamura, M. Suzuki, F. Inagaki, T. Yamakawa, and A. Suzuki, **J. Biochem.**, 1987, **101**, 825

12 S.W. Homans, R.A. Dwek, T.W. Rademacher, **Biochemistry**, 1987, **26**, 6553

13 S.W. Homans, R.A. Dwek, and T.W. Rademacher, **Biochemistry**, 1987, **26**, 6571

14 S.W. Homans, A. Pastore, R.A. Dwek, and T.W. Rademacher, **Biochemistry**, 1987, **26**, 6649

15 D.L. Fernandes, D.Phil. Thesis, 1986, Oxford University.

16 A. Imberty, H. Chanzy, S. Perez, A. Buleon, and V. Tran, **J. Mol. Biol.**, 1988, **201**, 365

17 C.E. Sanson, E.D.T. Atkins, and C. Upstill in '**Gums and Stabilisers for the Food Industry, 3**' (Eds. G.O. Phillips, D.J. Wedlock and P.A. Williams), Elsevier, London, 1986, 565

18 R.J.P. Williams, **J. Theoret. Biol.**, 1986, **121**, 1

19 A. Allen, **Trends Biochem. Sci.**, 1983, **8**, 169

20 K. Olden, B.A. Bernard, M.J. Humphries, T.K. Yeo, S.L. White, S.A. Newton, H.C. Bauer, J.B. Parent, **Trends Biochem. Sci.**, 1985, **10**, 78

21 R.J. Leatherbarrow, T.W. Rademacher, R.A. Dwek, J.M. Woof, A. Clark, D.R. Burton, N. Richardson, A. Feinstein, **Mol. Immunol.**, 1985, **22**, 407

22 M.A.J. Ferguson, R.A. Dwek, S.W. Homans, T.W. Rademacher in 'Host Parasite Cellular and Molecular Interactions in Protozoal Infections', (Ed., K.–P. Chang and D. Snary), 1987, Springer–Verlag

23 E. Osterlund, I. Eronin, K. Osterlund, M. Vuento, **Biochem.**, 1985, **24**, 2661

24 K.M. Yamada, **Ann. Rev. Biochem.**, 1983, **52**, 761

25 G.M. Edelman, **Ann. Rev. Biochem.**, 1985, **54**, 135

26 M. Gallantin, T. St. John, M. Siegelman, R. Reichert, E.C. Butcher, I.L. Weissman, **Cell**, 1986, **14**, 673

27 A.R. Poole, **Biochem. J.**, 1986, **236**, 1

28 C.L. Reading, in '**The Biology of Glycoproteins**' (Ed., R.J. Ivatt), 1984, Plenum Press, New York and London

29 G. Ashwell and J. Harford, **Ann. Rev. Biochem.**, 1982, 51, 531

30 T. Feizi and R.A. Childs, **Trends Biochem. Sci.**, 1985, **10**, 24

31 J.T. Gallagher, M. Lyon, and W.P. Steward, **Biochem. J.**, 1986, 236, 313

32 R. Schauer, **Trends Biochem. Sci.**, 1985, **10**, 357

33 F.O. Calvo and R.J. Ryan, **Biochem.** 1985, **24**, 1953

34 M. Fukuda and M.N. Fukuda, in 'The Biology of Glycoproteins' (Ed., R.J.
    Ivatt), 1984, Plenum Press, New York and London

35 W.J. Lennarz, **Trends Biochem. Sci.**, 1985, **10**, 248

36 K. Yamashita, Y. Tachibana, T. Ohkura, and A. Kobata, **J. Biol. Chem.**,
    1985, **260**, 3963

37 R.J. Ivatt, in 'The Biology of Glycoproteins' (Ed., R.J. Ivatt), 1984,
    Plenum Press, New York and London

38 J. Finne, **Trends Biochem. Sci.**, 1985, **10**, 129

39 S. Bozzaro, **Cell Differentiation**, 1985, **17**, 67

40 R.B. Parekh, R.A. Dwek, B.J. Sutton, D.L. Fernandes, A. Leung, D. Stan-
    worth, T.W. Rademacher, T. Mizuochi, T. Taniguchi, K. Matsuta, F.
    Takeuchi, Y. Nagano, T. Miyamoto, and A. Kobata, **Nature**, 1985, **316**,
    452

41 H. Lodish and N. King, **J. Cell. Biol.**, 1984, **98**, 1720

42 K. Yamashita, A. Hitoi, N. Tateishi, T. Higashi, Y. Sakomoto, A. Kobata,
    Arch. Biochem. Biophys., 1985, **240**, 573

43 T.W. Rademacher, S.W. Homans, R.B. Parekh, and R.A. Dwek, **Biochem.
    Soc. Symp**, 1986, **51**, 131

44 S.J. Swiedler, J.H. Freed, A.L. Tarentino, T.H. Plummer Jr., G.W. Hart, **J.
    Biol. Chem.**, 1985, **260**, 4046

45 R.B. Parekh, A.G.D. Tse, R.A. Dwek, A.F. Williams, T.W. Rademacher,
    EMBO J., 1987, **6**, 1233

46 W.D. Comper and T.C. Laurent, **Physiol. Revs.**, 1978, **58**, 255

47 A.G. Ogston in **Chemistry and Molecular Biology of the Intracellular Matrix** (Ed. E.A. Balaz), Academic Press, London, 1970, pp. 1231–1240

48 A. Katchalsky in **Progress in Biophysics and Biophysical Chemistry** (Ed. J.A. Butler and J.T. Randall), Pergamon Press, London, 1954, **Vol. 4**, pp. 1–59

49 M. Doi and S.F. Edwards, **The Theory of Polymer Dynamics**, Clarendon Press, Oxford, 1986.

50 J.E. Scott, F. Heatley, M.N. Jones, A. Wilkinson, and A.H. Olaveson, **Europ. J. Biochem.**, 1983, **130**, 491

51 G.S. Manning, **J. Chem. Phys.**, 1969, **51**, 924

52 P.S. Low, **Biochim. Biophys. Acta.**, 1986, **864**, 145

53 M.R. Egmond, R.J.P. Williams, E.J. Welsh, and D.A. Rees, **Europ. J. Biochem.**, 1979, **97**, 73

54 L.–A. Fransson, **Trends Biochem. Sci.**, 1987, **12**, 406

55 A.F. Williams and J. Gagnon, **Science**, 1982, **216**, 696

56 S.A. Jones, D.M. Goodall, A.N. Cutler, and I.T. Norton, **Europ. Biophys. J.**, 1987, **15**, 185

57 P. Gettins, J. Boyd, C.P.J. Glaudemans, M. Potter, and R.A. Dwek, **Biochemistry**, 1981, **20**, 7463.

58 J.T. Gallagher, **Bioscience Reports**, 1984, **4**, 621

59 E. Rowatt and R.J.P. Williams, **Biochem. J.**, 1987, **245**, 641–647

60 E.H. Eylar, **J. Theoret. Biol.**, 1966, **10**, 89

61 P. Albersheim and A.G. Darvill, **Glycoconjugates Proc. VIIIth Int. Symp.**, 1985, **2**, 511

19

Magnetic Resonance Studies of Molecular Dynamics at Membrane Surfaces

By A. Watts

BIOCHEMISTRY DEPARTMENT, UNIVERSITY OF OXFORD, SOUTH PARKS ROAD, OXFORD, OXI 3QU, U.K.

1 INTRODUCTION

The types and range of motions which occur within biological membranes are probably greater and more complex than within any other macromolecular assembly. So complex are these motions, that their comprehensive description is still not clear and attempts to deduce such a description poses a very significant technical challenge. Most importantly, considerable effort still has to be made to understand the relationship between membrane function and the dynamic events which occur within membranes, as well as how such dynamics can either be controlled by, or themselves control, membrane function. For such an exercise, physical methods such as spectroscopy, especially magnetic resonance spectroscopy, need to be correlated with biochemical approaches (and more recently genetic tools) to characterize one or more functions. To achieve this correlation between membrane dynamics and function is often the most difficult and neglected aspect of such investigations, as shown by simply a cursory review of the recent literature where such attempts have been made, which is all too infrequently.

A number of reviews exist describing comprehensively the use of magnetic resonance to study membrane dynamics, mainly those which take place in the membrane hydrophobic core[1,2,3,4,5,6]. Therefore this brief review will be restricted to our more recent work in which we have concentrated our efforts on understanding membrane surfaces.

Membranes contain a very high degree of chemical heterogeneity at their polar-apolar interface, it is therefore highly likely that this is where molecularly specific interactions could take place and potentially control biochemical function. Furthermore, membrane surfaces are the initial sites of contact with other cells, metabolites, ions, hormones, drugs, water and so on. In addition, cell fusion, cell division and any related events which involve structural destabilization of the membrane bilayer, may well be initiated at the cell surface. Indeed, it has been suggested that certain phospholipids, by virtue of their head group size, may be instrumental in promoting macromolecular rearrangements leading to the initiation of the rupture of bilayer structure which needs to take place if they are to fuse with other membranes[7].

Despite the important involvement of the membrane head group region in both structural and functional consequences for a cell, this region of the bilayer membrane has received rather little attention and hence scant molecular information is available about the molecular details of membrane surfaces and even less about the mutual interactions between different membrane

components. This discussion will therefore concentrate on
the associations between lipids and proteins which may
occur at the membrane surface.

2 MAGNETIC RESONANCE TO STUDY MEMBRANES

The molecular motions of all membrane components are
highly anisotropic. The two dimensional nature of the
membrane bilayer means that molecules can rotate around
their long molecular axis parallel to the bilayer normal
in a relatively unhindered way and at the same time
diffuse laterally within the plane of the bilayer.

This molecular anisotropy produces problems in the
use of many spectroscopic methods, for example proton and
carbon-13 NMR. However, spectroscopic methods which can
report on this molecular anisotropy due to their own
inherent anisotropic interaction with an applied
radiation, have proved very useful in describing the
molecular motions within biomembranes. Notably, electron
spin resonance (ESR) studies of stable free radical
nitroxide groups co-valently attached to phospholipids
and proteins, have resolved almost every known feature of
molecular motion in membranes[1]. Nuclear magnetic resonance
(NMR) studies, by virtue of the problems encountered in
resolving anisotropic molecular interactions from protons
and carbon-13[6], have been confined in the main to studies
of phosphorus-31 and specifically deuterated membrane
components[2,4,6,8]. In all these studies, an appreciation
of the dynamic time-scale of the spectroscopic method is
required before deductions about molecular information
can be made.

Rates of lipid molecular motion in membranes from magnetic resonance

It is possible to make some estimate of whether a particular motion is fast or slow with respect to a magnetic resonance spectroscopic property. To a first approximation, the molecular correlation time, τ_c, (or rate $= 1/\tau_c$) should be compared with the magnetic resonance property, or rate of change of the magnetic resonance property, being considered, expressed in frequency units, ω. Then if $\omega^2 \gg \tau_c^{-2}$, the molecular motion is fast and conversely if $\omega^2 \ll \tau_c^{-2}$, the molecular motion is slow on the time-scale of the magnetic resonance method.

Anisotropic magnetic properties which can be considered for membrane studies are the averaging of the anisotropy of deuterium NMR quadrupolar interactions ($2\pi\omega \sim 125$kHz for a CD bond) and nitroxide spin-label electron-^{14}N hyperfine interactions ($2\pi\omega \sim 2.5$mT $= 70$MHz), both of which can be averaged through molecular motion and can give information about the order of the system if the rate of molecular motion is fast with respect to ω (see below).

From determinations of deuterium spin-lattice (T_1) relaxation ($2\pi\omega \sim$ GHz) and spin-label spin-lattice (T_1) relaxation times ($2\pi\omega \sim$ MHz), it is possible to estimate the *rates* of molecular motion for the labelled group. Correlation times, τ_c, may be deduced from comparison of magnetic resonance spectra produced by isotopically tumbling labelled molecules with known radius, r, and in a solution of viscosity, η, (calculated from $\tau_c =$

$4\pi\eta r^3/3kT$) as, for example, in saturation transfer ESR of nitroxide labelled proteins[9]. For deuterium NMR, spin-lattice (T_1) relaxation times can measured and be used to give correlation times, as shown for deuterated lipids with low degrees of anisotropy in bilayers[10] where $1/T_1$ is proportional to τ_c.

Other magnetic properties, such as spin-spin (T_2) relaxation which occurs through molecular collision and thus direct proximal magnetic interactions between paramagnetic species or nuclei, have also been exploited in membrane systems, for example to determine lipid lateral diffusion rates. Here, the rate of relaxation, which takes place more slowly than for T_1, is typically sensitive to motions in the microsecond - millisecond time range for nitroxides and deuterons[11].

Figure 1 attempts to represent the range of motional *rates* of some molecular components which exist within biomembranes. From the above it can be seen that to determine the rate of any specific molecular motion, requires that the method employed to determine that motion must produce spectral sensitivity within the time-frame in which the motion of interest takes place. As an illustrative example of this, it would not be informative to attempt a determination of the rate of methyl group rotation ($\tau_r \sim 10^{-11}$s) on a protein side group or at the terminus of an acyl chain, by saturation transfer ESR whose sensitivity to molecular motion is in the microsecond time-range.

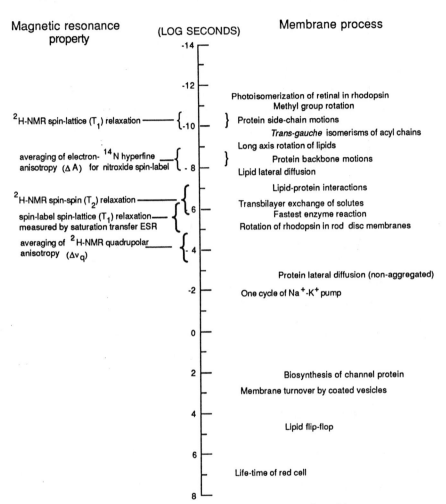

FIGURE 1: Some features of deuterium NMR and nitroxide spin-label ESR methods as applied to the study of membranes (left) and a selection of some molecular events in membranes for which the magnetic resonance methods can be applied and have motional sensitivity (right).

Further magnetic resonance features and properties
are also shown in Figure 1 with some indication of the
different membrane properties where motional sensitivity
overlap potentially occurs. Having stated this, it is not
always experimentally so straight-forward to get the
required information. In addition, some magnetic
resonance properties are not well described or understood
theoretically and they may be empirically based and yield
semi-quantitative information.

Particularly important is the case for molecular
exchange where a molecule can sample different
environments, say a protein interface and the bulk
bilayer. Depending upon the exchange frequency ($2\pi\omega_{ex}$ or
$1/\tau_{ex}$), either single averaged spectra, or two well
resolved spectra, characteristic of the two
distinguishable environments, will be observed for fast
and slow molecular exchange respectively as illustrated
in Figure 2. Only for the case of intermediate exchange
can the rate be directly determined, with limits given in
the extremes of exchange; this is explained later.

Order and amplitude of molecular motion

The *amplitude* of motion for a molecular bond or
group of interest, can vary between the limits of
rigidity ($\vartheta = 0°$) and isotropic motion ($\vartheta = 90°$) with
respect to a given reference, usually the normal to the
membrane bilayer plane. A tyrosine on a protein for
example, could undergo either full rotation or flip
between defined angular limits, depending upon its local
environment. Similarly lipid head groups could be
oriented rigidly at some defined angle to the membrane

normal or oscillate between a parallel and perpendicular orientation to the membrane normal.

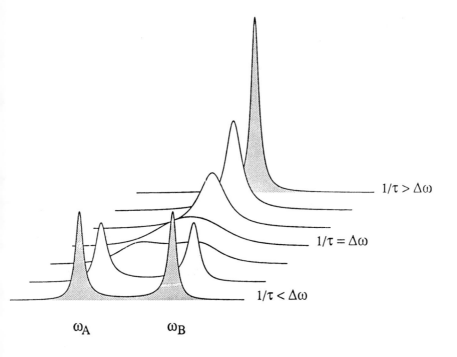

FIGURE 2: Representation of the effect of chemical exchange between two environments, A and B, in which the label has a frequency of ω_A and ω_B in the absence of exchange, $1/\tau \ll \Delta\omega$, at intermediate exchange, $1/\tau \sim \Delta\omega$ and fast exchange, $1/\tau \gg \Delta\omega$ where τ is the correlation time for the exchange process.

It is this motional molecular anisotropy which is most successfully exploited in magnetic resonance studies of membrane *order*. If the molecular motion of interest is fast (see above), then its average *amplitude*, ϑ, can be determined from $S = 1/2(3\langle\cos^2\vartheta\rangle - 1)$; the brackets ($\langle\ \rangle$) require that such a determination is only made for fast motional rates. (Slow motion distorts the spectral line

positions which are required to make order parameter
determinations). Then the molecular order parameter, S,
can vary between 1 ($\vartheta = 90°$) and $-1/2$ ($\vartheta = 0°$).
Experimentally, the spectral determination is made simply
from the ratio of the observed anisotropy/maximum
anisotropy and clearly varies between 0 and 1.
Quantitatively, for nitroxides the rigid limit maximum
anisotropy in the electron-^{14}N coupling constants are
given by the crystal values ($A_{xx,yy} - A_{zz}$) ~ 2.5mT[12] which
is ~ 0.7GHz in frequency units, and for deuterium NMR for
a CD_2 group the maximum anisotropy is $3/4(e^2qQ/h)$
~125kHz[2].

It can now be seen how the spectroscopic time-scale
is defined with respect to the molecular motional *rates*,
with fast molecular motion being of a rate sufficient to
average out the anisotropy detected in the magnetic
resonance spectrum.

For the special case of the molecular tensor being
at a value of $\vartheta = 54° 55'$, the magic angle, all spectral
anisotropic interactions are collapsed, even though the
molecular anisotropy is clearly rather significant. Since
it is the anisotropy of the magnetic interactions which
can severely restrict the use of some sensitive nuclei,
such as protons and ^{13}C for membrane study (because of
line-broadening and loss of high resolution information),
the potential now exists to resolve narrow NMR spectral
lines (due to the collapsed spectral anisotropy) from
macromolecules if they can be rotated at this magic angle
with respect to the applied magnetic field; the speed of
rotation must, however, be at a speed fast enough
(usually 5 - 20kHz) to average out the spectral

anisotropy (magic angle sample spinning, MASS). Practical problems which exist when handling membrane samples are being resolved in some circumstances.

3 LIPID HEADGROUP DYNAMICS

The polar head groups of phospholipids in bilayers of single phospholipid types display a relatively high degree of molecular order. This is evident from the deuterium NMR spectra of specifically head group deuterated lipids. These groups give rise to characteristic spherically averaged powder patterns typical of CD-groups in an ordered array of molecules rapidly moving about and around their long molecular axis. The quadrupole splitting, $\Delta\nu_q$, is now sensitive to the angle defining the amplitude of motion (θ) about an axis which may or may not be coincident with the long molecular axis. Thus, any detected changes in measured quadrupole splittings can reflect changes in the amplitude of motion of a CD-group at a membrane surface, the conformation of that head group, or both. A simple decision on which of the changes in molecular terms is causing the spectral change, can rarely be made without added information, for example, neutron scattering from deuterated phospholipids in bilayers. Thus the non-perturbing nature and high degree of sensitivity to surface interactions, make the method very powerful.

As a specific example of this sensitivity, the D-NMR spectrum from bilayers of phosphatidyl glycerol with deuterons in the glycerol head group, shows different quadrupole splittings for the optically active carbon-2 of the glycerols in the phospholipid, both for the backbone glycerol (which is also reflected in the head group orientation) and for the head group glycerol itself. This arises through the magnetic inequivalence of the various deuterons with respect to the bilayer caused by the chiral centres, in particular of the α-CD$_2$-group whose resonance is split into four quadrupole splittings for the phospholipid with a *sn*-glycerol backbone and *rac*-glycerol head group[13].

$$R_1\text{-COO} - CH_2 \qquad \text{Phosphatidyl glycerol}$$
$$R_2\text{-COO} - \overset{*}{C}H \qquad O$$
$$CH_2\text{-O-}\overset{\|}{P}\text{-O-}CD_2\text{-}\overset{*}{C}D(OH)\text{-}CD_2OH$$
$$\underset{O^-}{\big|} \qquad \alpha \quad \beta \qquad \gamma$$

The deuterium NMR method therefore has the capability of being used to determine the chirality of some specific deuterated compounds.

Lipid-lipid Interactions and membrane surface electrostatics

Lipids interact with each other at bilayer surfaces[14,15]. They have varying degrees of order and the polar head groups occupy varying volumes at the bilayer surface defined by their chemistry, with the overall area swept at the surface for each molecule being determined by the acyl chain properties. Thus, the relatively large volume occupied by the choline group (-N$^+$(CH$_3$)$_3$) of phosphatidylcholine infers an average tilt on the phosphocholine (-P(O$_2$$^-$)-O-CH$_2$-CH$_2$-N$^+$(CH$_3$)$_3$) moiety with respect to the bilayer plane of 35°, whereas the polar groups of most other phospholipids are virtually parallel to the bilayer plane[16]. These groups rotate around their long molecular axis, as does the whole molecule within the bilayer, with rotational τ_c values of 10^{-9}s as shown by spin-label ESR studies[1].

Intermolecular interactions between zwitterionic phospholipids appear to be rather weak. Phosphatidylcholines and phosphatidylethanolamines, two major eukaryotic plasma membrane phospholipids, can be mixed over a wide concentration range with little perturbation of the conformation or amplitude of motion of one lipid head group on the other[14,15]. It was suggested that such PE head groups are rather resistant to conformational perturbations by other lipids. X-ray studies of phospholipid crystals (admittedly at low hydration and in non-aqueous solvents) have suggested that PE molecules have a capability to form hydrogen bonds with other phosphatidylethanolamines[16], without the need for water molecules. Since it is thought that water molecules are required for similar non-covalent bonding at the membrane surface for other phospholipids, phosphatidylethanolamines may be more resistant to conformational changes than other lipids since these *intra*-molecular hydrogen bonds will be much stronger than those mediated through water for other lipids. This raises the important question of whether phosphatidylethanolamines have a specific rôle in associating strongly with proteins, thereby sealing the lipid-protein interface. Functional evidence to support this suggestion, is so far not available, although some suggestions have been made from reconstitution experiments where phosphatidylethanolamines, with other lipids, seem necessary for full activity of a protein. In addition, it has been suggested that phosphatidylethanolamines are primary lipids involved in bilayer destabilization and have potential for alternative phase formation in biological membranes, as a direct result of their lower states of hydration and smaller polar head group than other lipids[7].

The *rate* of interconversion between different conformers of phospholipid head groups is usually fast (within the quadrupolar anisotropy averaging time-scale). Spin-lattice (T_1) relaxation times are typically in the 10 - 30ms time range. These relaxation times give reorientational correlation times of the CD-groups with respect to the applied field in the nanosecond range reflecting the rather fast motion of phospholipid head groups.

Despite this fast rate of motion, the activation energies for reorientation of the polar head group segments are relatively high (Figure 3), as determined from the temperature dependence of the spin-lattice relaxation times (Dempsey & Watts, unpublished data). In fact, the acyl chains have less restriction to their rotational reorientations than polar methylene segments, implying that surface non-covalent bonding is important in ordering the polar group.

By placing deuterons in the polar head group of bilayer phospholipids it has been possible to examine the nature of interactions which can perturb the conformation of phospholipid head groups[8]. For example, it has been shown that changes of electrostatics at bilayer surfaces do significantly perturb the conformation and/or orientation of phospholipid head groups. For example, in mixed bilayers of phosphatidylethanolamines and phosphatidylcholine molecules, changing the bulk aqueous pH is seen to be reflected in the measured quadrupolar splittings from deuterated head group segments, as shown in Figure 4[17]. Since any changes in the electrostatic shielding of deuterium nuclei are much smaller than the quadrupolar changes which reflect orientational changes in the CD-group, (any such changes would, in any case, simply be reflected in the field position of the centre of the spectrum), such electrostatic sensitivity is a direct result of conformational changes at the bilayer surface. Indeed, small orientational changes (a few degrees only) can produce readily detectable changes in the deuterium spectrum, reflecting the sensitivity of the method.

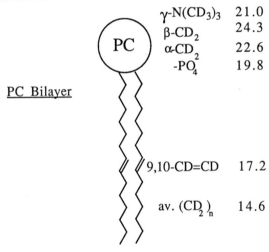

γ-N(CD$_3$)$_3$	21.0
β-CD$_2$	24.3
α-CD$_2$	22.6
-PO$_4$	19.8
9,10-CD=CD	17.2
av. (CD$_2$)$_n$	14.6

PC Bilayer

PC

FIGURE 3: Activation energies (kJ/mole) for fast motions of various segment of a phospholipid in hydrated bilayers from the temperature dependence of deuterium NMR spin-lattice (T$_1$) relaxation times.

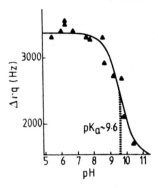

Figure 4: Measured deuterium NMR quadrupole splittings, $\Delta\nu_q$, for β-C^{2}H$_2$ of the polar head group of phosphatidyl ethanolamine, plotted as a function of the bulk pH of the dispersion of equimolar bilayers of phosphatidyl ethanolamine and phosphatidyl choline at 57°C (from ref 17).

Since one component D-NMR spectra are recorded from bilayers in the presence of ions and composed of lipid mixtures, then the exchange of ions (including protons) onto and off the bilayer surface is fast within the deuterium quadrupolar anisotropy averaging time scale. Thus, no lateral phase separation of lipids into areas rich in any one lipid type or different degree of lipid ionization, are produced such that any one area exists for longer than 10 - 100μs (the reciprocal of the quadrupolar splittings).

The demonstration that D-NMR can be used as sensors of electric charge in membranes[18] has been subsequently extended to investigating ionic interactions with bilayer surfaces, in particular in defining absorption isotherms for divalent ions with charged and zwitterionic bilayers. The relationship between measured quadrupolar splitting and surface potential has been suggested as being linear, with the relationship:

$$\Delta v_q \ (A) \ = \ \Delta v_q \ (o) \ + \ \chi_{c_A}$$

where Δv_q are the measured quadrupole splittings for the pure lipids bilayer (o) and in the presence of amphiphile (A) at a mole fraction of χ_{c_A}. In the examples studied to date, it appears that deuterated phospholipid head groups can report as *molecular sensors* to surface charge on membranes. If calibration of the experimentally determined deuterium NMR quadrupolar splitting with surface electric charge is possible, then here we have a means of directly probing the average surface charge density of a membrane surface, and its perturbation by, for example, oligosaccharides, proteins, ions, and other extra-membraneous and surface bound membrane components. The exchange of ions onto and off the bilayer surface is, as expected, fast on the quadrupolar averaging time-scale which ensures that the changes monitored experimentally are weighted means of the populations of different ionization states of the groups giving rise to the NMR spectral changes. Such groups are not the deuterated segments themselves; these would have freely exchanging deuterons. However, the sensitivity of the method is such that, for example with phosphatidyl ethanolamines, only the CD_2-group adjacent to the ionizable group reports on the ionization of the group whereas the other CD_2-group in the polar head group reports on its covalently attached nearest neighbour.

4   SELECTIVITY   AND   MOLECULAR   EXCHANGE   OF PHOSPHOLIPID-PROTEIN   INTERACTIONS

Magnetic resonance methods (especially nitroxide spin-label ESR and deuterium NMR) have been particularly fruitful in revealing the nature of phospholipid-protein interactions. Averaging of the spectral anisotropy by the motion of phospholipids has been found to be different for labels at the protein interface and in the bulk lipid phase. The resolution of this differential motion, due to the motion of the protein itself inferring a slower motion on neighbouring lipids when compared to lipids in the protein-free bulk phase, has been a key to the success of the methods. This differential motion reveals itself as two (or perhaps more)-site exchange between the two (or more) motionally distinct environments, and an understanding of how to analyze this exchange from the spectral features has been essential in obtaining molecular information about protein-lipid interactions.

Initial indications of direct phospholipid-protein interactions were provided by nitroxide ESR spin-label studies[19]. Here, the rate of motion for labels attached to the lipid acyl chains at the interface of a large integral protein have been found to be sufficiently different (by at least an order of magnitude) from lipid chains in the bulk lipid phase away from the protein interface such that the ESR spectrum recorded is two component in nature. Each component, from labels slowly exchanging at a frequency, v_{ex}, in the $0.1 - 1.0 \mu s$ range between the two motionally distinct environments, could be quantitated to give estimates for the fraction of lipids in either of the two membrane environments[5,21,22].

The key to this approach is the use of the exchange coupled, modified Bloch equation for transfer of spin magnetization between lipids at the protein interface and those in the bulk phase. Thus, if the magnetization associated with lipids (spin-labelled or deuterated) at the boundary of a protein is M_b, then:

$$\frac{dM_b}{dt} = -\tau_b^{-1}M_b + \tau_f^{-1}M_f$$

where t is the inverse of the probability of transfer per unit time of a labelled lipid from the protein interface, b, to the fluid bilayer, f. At equilibrium, $dM_b/dt = dM_f/dt = 0$. Including this assumption into the Bloch equations, some estimate of the exchange rates and stoichiometry of lipid-protein association can be made from computer simulation of the magnetic resonance spectra and fitting the frequency overlap (as shown in Figure 2) due to the exchange interactions (see ref 22 for a more complete description). Importantly, the rate of molecular exchange which can be determined, depends upon the spectroscopic technique being used. That is, the frequency difference, $\omega_A - \omega_B$ (Figure 2), is determined by the technique and its intrinsic time-scale (Figure 1) as described earlier. Therefore it is now possible to use different spectroscopic methods to identify different rates of molecular exchange.

In the ESR spin-label case, estimates for lipid-protein association rates and stoichiometries have been made for a small number of proteins (for example, interactions between rhodopsin and phosphatidyl choline[23],

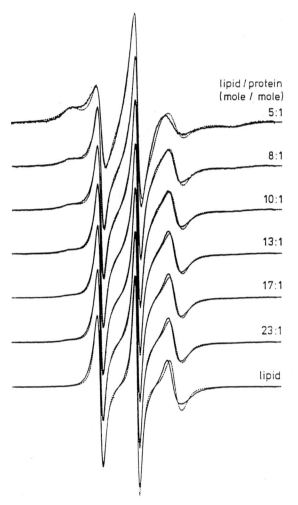

Figure 5: ESR spin-label spectra for an acyl chain labelled phosphatidyl choline in bilayer containing increasing concentrations of M13 bacteriophage coat protein (solid line) compared with the computer simulated ESR spectra (dotted line) using the exchange coupled modified Bloch equations to give a lipid-protein exchange rate for the lipids off the protein interface of ~ 5MHz. The broader spectral component increases in intensity relative to the total intensity linearly with the protein content in the bilayer (from ref 24).

the M13 bacteriophage coat protein and phosphatidyl choline[24], myelin proteolipid and phosphatidyl choline[25] assuming that only two motionally distinct environments exist in the membrane bilayer. Figure 5 shows a typical set of experimental and simulated nitroxide ESR spectra to demonstrate the goodness of fit with this approach for one particular system, the M13 coat protein in phosphatidyl choline bilayers. Typically, the lipid-protein exchange rate, ν_{ex}, determined using a labelled phosphatidyl choline, does not change significantly with protein concentration in the phosphatidyl choline bilayer.

TABLE 1. Stoichiometries of the Motionally Restricted Lipid Component in Various Lipid–Protein Systems, N_1^{exp}, and Estimates of the Number of First-Shell Lipids That Can Be Accomodated Around the Integral Proteins, N_1^{calc}*

Protein/membrane	MW × 10^{-3}	N_1^{exp} (mole/mole)	N_1^{calc} (mole/mole)
(Na^+, K^+)-ATPase-DOPC	314	63 ± 3	(~57–72)
(Na^+, K^+)-ATPase shark rectal gland	265	58 ± 4	(~57–72)
Cytochrome oxidase-DMPC	165	45 ± 4	40–45
Acetylcholine receptor-DOPC	250	40 ± 7	43
Ca^{2+}-ATPase–egg PC	115	22 ± 2	(~23–27)[a]
Sarcoplasmic reticulum/ Ca^{2+}-ATPase	115	24	(~23–27)[a]
Bovine rod outer segment disk/rhodopsin	39	25 ± 3	24 (±2)
Frog rod outer segment disk/rhodopsin	39	23 ± 2	(24)
Myelin-proteolipid apoprotein–DMPC	25	10 ± 2	(~10–12)[b]

DMPC, dimyristoyl phosphatidylcholine; DOPC, dioleoyl phosphatidylcholine; MW, molecular weight. For references to original ESR data and details of calculations of the number of first-shell lipids around the proteins, see Marsh (1985).

[a]Calculated assuming a dimer, with monomer radius 20 Å.
[b]Calculated assuming hexamer.

Adapted from ref 20.

Interestingly, the stoichiometry of phospholipids at the protein interface can give an indication of the hydrophobic interface of the protein and hence the protein diameter if some assumption about the cross-sectional shape is made, most conveniently a cylinder, Table 1. This can then help in the estimation of how much of the protein density is buried in the membrane and how much is extramembraneous[26,27].

The suggestion that specific lipid types could preferentially interact with integral proteins was demonstrated for cytochrome *c* oxidase using the ESR spin-label method[28]. Six types of phospholipids, with nitroxide spin-labels covalently attached to C14 of the *sn*-2 chain, were introduced into a complex of cytochrome *c* oxidase reconstituted into dimyristoyl phosphatidylcholine bilayers at a lipid to protein ratio of about 100:1. Four of the labels, the choline, ethanolamine, glycerol and serine phospholipid analogues all displayed similar degrees of association with the protein interface whereas the cardiolipin and phosphatidic acid analogues both showed a significantly enhanced association with the protein.

The degree of association could be interpreted in one of two ways. Either a larger number of sites at the protein-lipid interface were available to the more strongly selected lipids (with 55 lipids per protein for most of the labels and 59 and 77 for the phosphatidic acid and cardiolipin labels respectively) or a stronger association (with relative associations of 1 for most of the labels and 1.9 and 5.4 for the phosphatidic acid and cardiolipin labels respectively). A number of further examples have been demonstrated for other proteins since this first example of cytochrome *c* oxidase, as shown in Table 2. The important question still remains, however, is the elucidation of a structural selectivity of any functional significance with an integral protein. Although some examples do exist in the literature of lipid-protein dependence on activity, the parallel structural studies have frequently not been so conclusive, and *vice versa*[29].

TABLE 2. Order of Selectivity of Spin-Labeled Lipids for Association With Integral Membrane Proteins*

Protein	Order of lipid selectivity
Myelin proteolipid	SA > PA > CL ⁓ PS > PG ⁓ SM ≈ PC > PE
(Na⁺, K⁺)-ATPase	CL > PS ≈ SA ⁓ PA > PG ⁓ SM ≈ PC ⁓ PE
Cytochrome oxidase	CL > PA ≈ SA > PS ≈ PG ≈ SM ≈ PC ≈ PE
Acetylcholine receptor	SA > PA > PS ≈ PC ≈ PE
Ca²⁺-ATPase	CL > PS ≈ SA ⁓ PA ⁓ PG ⁓ SM ≈ PC ≈ PE
Rhodopsin	CL ≈ PA ≈ SA ≈ PS ≈ PG ≈ SM ≈ PC ≈ PE

*From Marsh, 1985. CL, cardiolipin; PA, phosphatidic acid; PS, phosphatidylserine; PG, phosphatidylglycerol; PC, phosphatidylcholine; PE, phosphatidylethanolamine; SM, sphingomyelin; SA, stearic acid.

From ref 20

In the case of slow to intermediate exchange, some indication of the rate of spin exchange can be obtained from spectral simulations. In most cases of lipid-protein interaction studies using deuterium NMR of labelled lipids, either little effect of the protein on the deuterated phospholipid NMR spectra has been observed[30] or alternatively the lipids were shown to be in fast exchange throughout the whole bilayer complex[8,31,32,33]. This latter observation is not inconsistent with the results from spin-label experiments described earlier. The lipid-protein molecular exchange rates are slow with the ESR method but fast with the NMR methods by virtue of the different motional rates required to average the spectral anisotropy, as described earlier[34].

The complementary use of deuterium NMR of deuterated lipids and nitroxide spin-label ESR in similar (if not the same) protein-lipid complex, therefore could give rather detailed dynamic information about the various exchange processes of the lipids in the complex. This has been done recently by us with the coat protein from the filamentous M13 bacteriophage. Typical spin-label ESR spectra for a nitroxide labelled phospholipid in a bilayer complex containing the coat protein at various lipid-protein mole ratio are shown in Figure 5 above, in which the contribution of broader ESR spectral component increases with protein content, as observed with many other large integral proteins in bilayers. The exception here, however, is that the protein is a single *trans*-membrane 50-amino acid chain and not a large integral protein. The reason for the motional restriction of the lipids is that aggregates of individual proteins are sufficiently large as to behave like a very large protein in restricting the motion of the lipids, both at the aggregate boundary and within the protein aggregates. When the same complex is probed with deuterated lipids and observed by deuterium NMR, the origin of a separate central spectral component is also observed (Figure 6), which is rather unusual, since in many cases studied to date, the lipids are seen to be in fast exchange by NMR with the lipid-protein interface as explained above.

How can such information be rationalized, especially since the value for v_{ex} for lipids at the coat protein aggregate interface, is 0.3kHz from simulations of the deuterium NMR spectra, but 5MHz from analysis of the spin-label spectra? Our suggestion is that lipids are

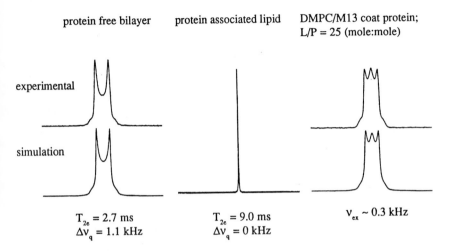

protein free bilayer   protein associated lipid   DMPC/M13 coat protein;
                                                                L/P = 25 (mole:mole)

experimental

simulation

T_{2e} = 2.7 ms T_{2e} = 9.0 ms ν_{ex} ~ 0.3 kHz
$\Delta\nu_q$ = 1.1 kHz $\Delta\nu_q$ = 0 kHz

Figure 6: Deuterium NMR spectra from protein-free bilayers and those containing M13 bacteriophage coat protein. The protein induces a central spectral component which increases as the protein content increases. The proportion of protein associated lipid has been deduced from computer subtractions and the exchange rate of lipids onto and off the protein aggregate interface is slower than 0.3kHz.

trapped inbetween large, probably co-linear, aggregates of the protein (van Gorkom, Horváth, Hemminga & Watts, in preparation). This means that some lipids are boundary lipids (in fast exchange by NMR but slow exchange by ESR) and some are trapped (in slow exchange by both NMR and ESR). Such aggregates may therefore be formed in the bacterial cell membrane before bacteriophage extrusion from its host. The protein aggregates are most probably linear, since 2D agregates of protein would decrease the boundary component as the aggregates increase in size, without increasing the amount of motionally restricted (trapped) lipids. Only with a linear growth of an aggregate can the amount of boundary lipid increase with such a small protein. As some confirmation of this, the stoichiometry for the motionally restricted lipid is different for the two methods, with more lipid contributing to the broadened lipid spectrum for ESR than for NMR. Here is a clear case therefore, where the use of two spectroscopic methods can give complementary information about the molecular dynamic processes in bilayer membranes.

6 CONCLUSION

It is becoming clear that with a deeper understanding and greater sophistication of the spectroscopic methods, the subtlety of molecular dynamics in membranes are becoming better resolved. Still, however, the functional aspects of these structural and dynamic phenomena need to be related more closely to the function of natural biological membranes. Extending current studies to mixed lipids, then with multi-component protein complexes, will help in this quest.

In addition, magnetic resonance methods are not the only ones which can be used to study membranes and as many other techniques as are available need to be used.

ACKNOWLEDGEMENTS

I wish to acknowledge the very fruitful collaborations with Marcus Hemminga and Derek Marsh, made possible through EEC grants ST2J-0088 and ST2-00368, as well as the support and valuable discussions with Leon van Gorkom, Chris Dempsey, Laszlo Horváth (funded on SERC travel grant grant GR/E/89070), Paul Spooner, Lee Fielding and Saira Malik. Also I wish to acknowledge SERC for research grants GR/E/69846, GR/E/56683, GR/F/13782.

REFERENCES

1. Marsh, D and Watts, A. Ch. 5 in: Liposomes: from physical   structure to therapeutic application. (Knight, C.G. ed) Elsevier, Amsterdam, 1981, Ch. 6, p139.
2. Seelig, J. & Seelig, A. Quart. Rev. Biophysics 13, 1980, p9 -61
3. Davies, J. H. Biochim. Biophys. Acta, 1983, 737, pp117-171
4. Bloom, M. & Smith, I.C.P. in: Progress in Protein-lipid Interactions, (Watts, A. & de Pont, J.J.H.H.M. eds) Elsevier, Amsterdam, 1985, Vol. 1, Ch 2, p61.
5. Devaux, P.F. & Seigneuret, M. Biochim. Biophys. Acta 1985, 822, 63-125
6. Deese, A. & Dratz, E. in: Progress in Protein-lipid Interactions, (Watts, A. & de Pont, J.J.H.H.M. eds) Elsevier, Amsterdam, 1986, Vol. 2, Ch 2, p45.
7. de Kruijff, B., Cullis, P.R., Verkleij, A.J., Hope, M.J., van Echtfeld, C.J.A., Taraschi, T.F., van Hoogevest, P., Killian, J.A., Rietveld, A. & van der Steen, A.T.M. in: Progress in Protein-ipid Interactions, (Watts, A. & de Pont, J.J.H.H.M. eds) Elsevier, Amsterdam, 1985, Vol. 1, Ch   3, p89.
8. Watts, A. J. Biomembranes and Bioenergetics, 1987b, 19, pp625-653
9. Thomas, D.D., Dalton, L. & Hyde, J. J. Chem. Phys. 1976, 65, 3006
10. Brown. M. F., Seelig, J. & Haberlen, U. J. Chem. Phys., 1979, 70, 5045-5053
11. Träuble, H. & Sackman, E. J. Chem. Amer. Soc. 1972, 94, 4499-4510
12. Jost, P.C., Griffith, O.H., in: Spin Labelling, (Berliner, L.J. ed) Academic Press, New York, 1976, Vol. 1.

13. Wohlegemuth, R., Waespe-Sarcevic, N. & Seelig, J. Biochemistry, 1980, 19, 3315-3321
14. Sixl, F. & Watts, A. Biochemistry, 1982, 21, 6446-6452
15. Sixl, F. & Watts, A. Proc. Natl. Acad. Sci. USA 1983, 80, 1613-1615
16. Hauser, H., Pascher, I., Pearson, R.H., and Sundrell, S. Biochim. Biophys. Acta, 1981, 650, 21-51
17. Watts, A. & Poile, T.W. Biochim. Biophys. Acta 1986, 861, 368
18. Seelig, J., Macdonald, P.M. & Scherer. P. G. Biochemistry, 1987, 26, 7535-7541
19. Jost, P.C., Griffith, O.H., Capaldi, R.A., & Vanderlooi, G. Proc. Natl. Acad. Sci. USA. 70, 1973, 480-484
20. Marsh, D. & Watts, A. in Adv. in Membrane Fluidity, Vol 2,(Aloia, R., Curtain, C.C. & Gordon, L.M. eds) Alan Liss, New York, 1988, Vol. 2,Ch. 9, 163.
21. Marsh, D. and Watts, A. Ch.2 in: Lipid Protein Interactions, Vol. 2. (Jost, P.C. & Griffith, O.H. eds) Wiley Interscience, New York, 1982, Vol. 1, Ch. 2, p53.
22. Marsh, D. in: Progress in Protein-lipid Interactions, (Watts, A. & de Pont, J.J.H.H.M. eds) Elsevier, Amsterdam, 1985, Vol. 1, Ch 4, p143.
23. Ryba, N.J.P., Horváth, L.I., Watts, A. & Marsh, D. Biochemistry 1987, 25, 4818-4825
24. Wolfs, C.J.A.M., Horváth, L.I., Marsh, D., Watts, A. Hemminga, M.A. Biochemistry, 1989, (submitted)
25. Horváth, L.I., Brophy, P.J. & Marsh, D. Biochemistry, 1988, 27, 46-54
26. Watts, A., Volotovski, I.D., & Marsh, D. Biochemistry, 1979, 18, 5006-5013
27. Knowles, P.F., Watts, A., & Marsh, D. Biochemistry, 1979, 18, 4480-4487
28. Knowles, P.F., Watts, A., & Marsh, D. Biochemistry, 1981, 21, 5888-5894
29. Watts, A. & de Pont, J.J.H.H.M. eds of Progress in Protein-lipid Interactions, Vols. 1 & 2, Elsevier, Amsterdam, 1985,1986
30. Tamm, L. & Seelig, J. Biochemistry, 1983, 22, 1474-1483
30. Watts, A. in: Membrane Receptors, Dynamics and Energetics,(Wirtz, K.A.W. ed) Plenum Press, New York-London, 1987a,   p329 .
31. Ryba, N.J.P., Dempsey, C.E., & Watts, A. Biochemistry, 1986, 25, 4818-4825

32. Dempsey, C.E. Ryba, N.J.P., & Watts, A.
    Biochemistry, 1986, 25, 2180-2187
33. Sixl, F., Brophy, P.J., & Watts, A. Biochemistry,
    1984, 23, 2032-2039
34. Watts, A. Nature 1981, 294, 512.

20

Dynamic Properties of Model Membranes by Quasi-elastic Light Scattering

By J. C. Earnshaw, R. C. McGivern, and G. E. Crawford*

DEPARTMENT OF PURE AND APPLIED PHYSICS, THE QUEEN'S UNIVERSITY OF BELFAST,
BELFAST BT7 1NN, NORTHERN IRELAND.

1 INTRODUCTION

Biomembranes are complex molecular systems within which dynamic processes and fluctuations are all-pervasive, causing or modulating biologically significant membrane processes.[1] Examples of such influences include the effects of dilational fluctuations of the lipid matrix upon enzyme activity, of 'bobbing up and down' motions of proteins and lipids on permeability and of thickness fluctuations upon membrane fusion and similar processes. Despite this significance the study of such dynamic membrane processes is comparatively novel.

Biomembranes can be regarded as interfaces between two fluids, and as for any fluid interface, their flatness and homogeneity will be destroyed by fluctuations.[2] Light is scattered by fluctuations: in a uniform medium of homogeneous dielectric constant a light beam propagates without disturbance. In a nonuniform medium light will be scattered, the angular distribution of the scattered intensity being related to the spatial distribution of the nonuniformities. If these evolve with time, the intensity scattered to some point in space will be time dependent. Here we are concerned with light scattered by fluctuations in molecular films. Observation of the autocorrelation function at a particular scattering vector yields the temporal evolution of particular spatial components of the fluctuations within the film.

Model Membrane Systems

The chemical and physical heterogeneity of real biomembranes poses very real problems for the interpretation of experimental data from novel techniques. Various chemically and physically more homogeneous model systems, which retain many of the dynamic aspects inherent in biomembranes, have been devised. The physical uniformity of such controlled systems makes it easier to design experiments to detect particular modes of fluctuation. Similarly, control of the chemical composition permits investigation of the role of molecular structure upon the observed fluctuations.

---

*Present adress: BKS Surveys Ltd., Ballycairn Road, Coleraine, Northern Ireland

In these systems the membrane can be considered either as a supra-molecular structure within which molecular motions give rise to the observed dynamic behaviour, or alternatively as an interfacial film characterised by various macroscopic properties. The film fluctuations are governed by molecular interactions, expressed macroscopically as elastic moduli and viscosities of the membrane. Presumably the two contrasting pictures must converge, but to date theoretical approaches to the molecular nature of membranes have not predicted their viscoelastic properties. Lacking knowledge of the detailed molecular mechanisms involved, the fluctuations of interest must be described in terms of various modes (e.g. membrane bending and squeezing or molecular splay, twist and order modes), governed by corresponding viscoelastic moduli.[2,3] This macroscopic model involves membrane properties defined in the hydrodynamic limit.

Taken together, the various fluctuations mentioned cause the so-called 'fluidity' of membranes. Unfortunately, like 'micro-viscosity', this concept is rather ill-defined, and does not readily lend itself to analysis of effects upon real membrane processes. However, specific modes of membrane or molecular motion, governed by well-defined moduli, can be analysed fully, motivating physical experiments to probe them.

The various membrane models display certain significant differences. Thus the lipid molecules are subject to different constraints which affect the membrane behaviour: e.g. interbilayer forces due to membrane bending in multi-lamellar dispersions[4] and topological constraints on molecular tilt in vesicles[5] (precluding the pretransition found in other model systems). Similarly, some model systems can show different fluctuations: e.g. flaccid (tension-free) vesicles can display bending modes[6] which are suppressed in bilayer lipid membranes (BLM, or black lipid films) in which the effective bending modulus (Kq^2) is much less than the bilayer tension. The light scattering studies discussed here will largely concern BLM and monolayers at the air-water interface.

2 THEORETICAL BACKGROUND

We treat membranes as interfaces between fluids, neglecting for the moment their molecular constitution. Fluctuations in such membranes scatter light. As usual we analyse such fluctuations in terms of independent hydrodynamic modes of the system. There will be associated molecular motions, as well as possible modes of molecular motion which are separate from the membrane modes.

Rather basic considerations[3] suggest that an interface may possess up to five separate viscoelastic moduli, each associated with a well defined mode of motion (Figure 1). There is no unique way to express these modes, so we follow Goodrich:[3] the interface may be subject to both shear and dilational stress, each acting both within and perpendicular to the membrane plane, as well as slip in the interfacial plane. Fortunately not all of the associated motions contribute to the light scattering. This is not to deny the significance of those modes which do not scatter light, but simply to exclude them from the present considerations. Slip has usually been neglected, and for an isotropic membrane the tangential shear waves decouple from the other modes (and cause no change in dielec-

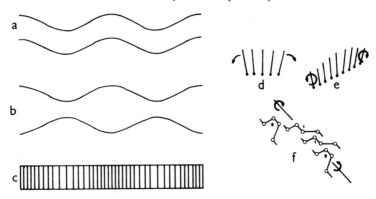

Figure 1: Membrane modes which can scatter light: a) capillary, b) thickness, c) compression, d) splay, e) twist and f) ordering.

tric constant of the film).[2] Dilation normal to the membrane causes thickness fluctuations, but these are restricted to wavenumbers far above experimentally accessible values.[7]

The hydrodynamic modes which may concern us in light scattering are then transverse shear and in-plane compression. The intensity scattered by the transverse shear (or capillary) waves greatly exceeds that due to the longitudinal compression waves,[8] so we concentrate upon the former. At a 'symmetric' fluid-fluid interface (identical fluids on either side, e.g. BLM) these two modes decouple,[2] whereas they are coupled for a monolayer at the water-air surface. Thus in the monolayer case the capillary waves are influenced by both of the relevant moduli, while for a BLM only the transverse shear modulus is involved.

The transverse displacement of the membrane from its equilibrium plane ($z = 0$) can be written

$$\zeta(\vec{r}, \tau) = \zeta_0 e^{i(\vec{q} \cdot \vec{r} + \omega \tau)} \tag{1}$$

where $\omega = \omega_0 + i\Gamma$. Experimentally, fluctuations of given wavenumber are selected by the scattering geometry, the frequency ω being measured. This frequency is related to the wavenumber (q), and to the material properties of the system, by the dispersion equation.[2] Apart from the densities and viscosities of the ambient fluids, the relevant properties are the transverse shear and in-plane dilational moduli of the membrane (respectively γ and ε). These are familiar in the equilibrium ($\omega \to 0$) limit: the transverse shear modulus as the interfacial tension and the dilational modulus as the inverse of the film compressibility. Associated with the two elastic moduli are interfacial viscosities which allow for dissipative processes. The moduli are expanded as response functions:

$$\gamma = \gamma_0 + i\omega\gamma' \tag{2}$$

$$\varepsilon = \varepsilon_0 + i\omega\varepsilon'. \tag{3}$$

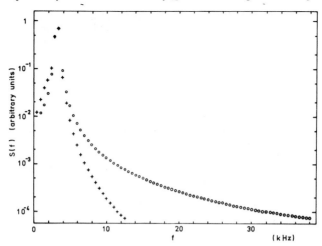

Figure 2: The spectrum of scattered light ($\times$) at $q = 180$ cm^{-1} for a surface of $\gamma_0 = 72.75$ mN/m, $\gamma' = 0$, $\varepsilon_0 = 3.6$ mN/m and $\varepsilon' = 10^{-5}$ mN s/m, together with the best fit Lorentzian form ($\circ$).

Such viscosities were first postulated by Goodrich,[9] and later deduced from fundamental arguments by both Goodrich[10] and Baus.[11]

The several interfacial properties have rather different effects upon the light scattering from the capillary waves. They affect the dispersion of the waves, expressed via the dispersion relation. The propagation of the capillary waves is directly governed by the elastic modulus γ_0 (the tension). The transverse shear viscosity γ' represents an additional dissipative influence upon the waves, its effect increasing as q^3.[12] The dilational modulus ε directly influences the longitudinal or compression modes, which only affect the capillary waves (in the monolayer case) through the coupling already mentioned. The light scattering is thus relatively insensitive to ε, except in circumstances where the two modes resonate (at accessible q values this occurs for $\varepsilon_0/\gamma_0 \sim 0.16$), the resonance being strongly damped by non-zero values of ε'. It is worth noting that for monolayers at a fluid-fluid interface (e.g. oil-water) the effects of ε are much reduced.[13]

Spectrum of Scattered Light

The theoretical spectrum of the light scattered by thermally excited waves of given q, which is just the power spectrum of the waves, has been deduced using the fluctuation-dissipation theorem.[13] This spectrum is a function of both the interfacial moduli γ and ε (in the monolayer case). It is *approximately* Lorentzian in shape (Figure 2), the deviations from that form being systematic functions of the surface properties. This fact has been exploited in data analysis to directly extract these properties from the observed light scattering data.[14,15]

However the indirect effect of ε upon the spectrum implies that such estimates of this modulus must be rather poorly determined, compared to γ. This is especially true of the viscosity ε', where order of magnitude estimates may be all that are obtainable from experimental data of reasonable quality.[16]

Viscoelastic Relaxation

Either or both of the membrane moduli may exhibit viscoelastic relaxation, the low frequency viscosity transforming into a contribution to the elastic modulus at frequencies above the reciprocal of the relaxation time scale. There is as yet no theoretical basis for membrane viscoelastic relaxation, and so we restrict ourselves to two particularly simple models: the viscoelastic (or Voigt) solid, in which the elastic and viscous portions of the relevant modulus are independent of frequency, and the viscoelastic (or Maxwell) fluid, involving a single exponential relaxation of stress (more complex patterns of behaviour can be simulated by arbitrary combinations of such elements). Oscillatory stress is related to the strain via the complex modulus $G^* = G' + iG''$, where for the Maxwell fluid[17]

$$G' = G_e + G\frac{\omega^2\tau^2}{1 + \omega^2\tau^2} \tag{4}$$

$$G'' = G\frac{\omega\tau}{1 + \omega^2\tau^2} \tag{5}$$

where τ and G are the time constant and strength of the relaxation process respectively, while G_e is the equilibrium, low frequency, value of the elastic modulus. We can identify G' with γ_0, or ε_0, and G'' with $\omega\gamma'$, or $\omega\varepsilon'$ (c.f. Eqns. 2 and 3). If the viscoelastic moduli were indeed to display such frequency dependence their effects upon the capillary wave propagation would be modified.

Molecular Modes

As mentioned above, molecular modes exist within the membranes along with the hydrodynamic modes discussed thus far. These molecular motions can be resolved into modes of splay, twist and ordering (c.f. Fig.1).[18] Two of these may couple to the hydrodynamic modes: splay with the capillary waves and ordering with the longitudinal waves. The molecular modes should be observable in depolarised scattering, but to date the only evidence from light scattering experiments has come from one study of BLM in which anomalies in the scattered intensities suggested the presence of splay modes which were strongly coupled to the capillary waves. However we note that these molecular splay modes are just the bending modes which have been observed in the fluctuations of flaccid vesicles.[6] The significance of these modes is suppressed for both BLM and monolayers by the membrane tension. These bending modes are also thought to cause the flicker phenomenon in red blood cells,[19] and the shape fluctuations of flaccid (tension-free) vesicles.[6] Recent light scattering experiments on single erythrocytes[20] are less well understood, although the signal appears definitely to be membrane-associated. These data may be explicable in terms of a q dependent bending modulus (K), as recently postulated.[21]

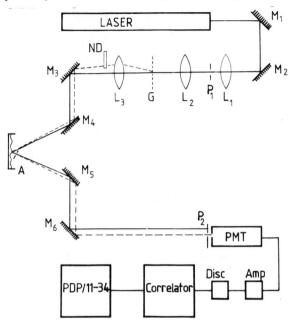

Figure 3: An outline sketch of our heterodyne surface spectrometer. The grating G acts as local oscillator.

3 EXPERIMENTAL CONSIDERATIONS

A typical experimental system,[12] involving autocorrelation of the photomultiplier output, is shown in Figure 3. Note that both propagating and overdamped modes may be observable, so that a heterodyne spectrometer is essential. It has become usual to use a 'weak' diffraction grating as a local oscillator. A lens images the grating at the membrane, the reference beams diffracted by the grating being thus caused to cross over at that point. These beams are much less intense than the zero-order beam, so that scattering from them is negligible. This arrangement ensures the spatial coherence of the reference and scattered light. By suitable adjustment of the scattered and reference beam intensities, the first order correlation function can be made to dominate the signal.

BLM and monolayers offer contrasting advantages. BLM are not easy to work with (large, planar, stable bilayers are required), but data analysis is easy as the two accessible membrane properties can be recovered from the complex frequency ω. Monolayers are experimentally simpler, but four membrane properties cannot uniquely be determined from the two observables ω_0 and Γ. However the measured autocorrelation function, which reflects the temporal evolution of the capillary waves, is just the Fourier transform of their power spectrum. The dependence of this spectrum upon the surface properties has been exploited, in the monolayer case, to determine the four membrane properties discussed above.[14,15] The only limitations of this relatively novel approach should be those inherent in the sensitivity of light scattering to the membrane properties.

It has proved possible to extract intensity information from the correlation data used to estimate the temporal behaviour of the capillary waves.[22] From the amplitude of the normalised correlation functions and the average photon count rate, relative values of the scattered and reference beam intensities (I_s and I_r, respectively) can be determined. Such relative values suffice for several purposes (see below); absolute intensities would require separate observations.

Recently we have found that it is possible to acquire heterodyne correlation functions extremely rapidly — much faster than for normal single photon correlation.[23] Such rapid data acquisition (e.g. experiments less than 10 ms) has obvious uses in the investigation of dynamic membrane processes, and has already contributed to studies of two-phase coexistence in monolayers.

Much of our experience has been with model membranes formed from glycerol monooleate (GMO). This lipid was originally chosen because of its ability to form large stable BLM, suitable for light scattering. Membranes of GMO also show a thermotropic transition at an experimentally accessible temperature — $T_t \sim 16.6°C$.[24] The present data all derive from this lipid.

4 RESULTS

Certain aspects of light scattering from model membrane systems have been reviewed previously,[25,26] including the first demonstration of non-zero values of the transverse shear viscosity γ', in 'solvent-free' BLM of GMO; the observation of the effects of BLM composition (lipid concentration, cholesterol content) upon the membrane properties; and various transitional studies of BLM and monolayers, in which molecular modes were found to contribute to the signal below T_t. One significant lesson has been that light scattering probes local membrane properties, averaged over the illuminated area.[27] This area, which is large on the molecular scale but smaller than the membrane, is, to some extent at least, subject to experimental control, providing useful insights in some circumstances. Here we will concentrate upon two relatively recent developments.

The ability to measure both the time dependence and the relative intensity of the scattered light in a single observation is rather powerful. This is illustrated by the demonstration of the thermal nature of the fluctuations which scatter the light in the cases of both BLM[22] and monolayers.[28] The scattered intensities (and hence the rms amplitudes of the capillary waves) varies as q^{-2} in both cases, as predicted. By varying the molecular packing for monolayers the expected dependence upon the tension, as γ_0^{-1}, has been verified (Figure 4), The observed variation of I_s with tension clearly shows that, as expected, the dilational modulus has essentially no effect upon the amplitudes of the thermally excited capillary waves: the equilibrium dilational modulus ε_0 increases from 0 to over 100 mN/m over the observed range of γ_0. If any viscoelastic relaxation were involved, the value of tension affecting the amplitudes of the capillary waves of particular q would be that appropriate to the wave frequency (c.f. Eqn. 4).

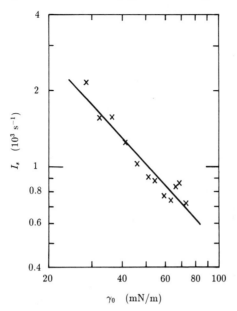

Figure 4: The tension dependence of scattered intensity from capillary waves ($q = 745.0$ cm^{-1}) on a GMO monolayer during film compression.

Viscoelastic Relaxation

The most exciting recent development is the observation of molecular effects upon viscoelastic relaxation in the two accessible surface moduli, γ and ε. Such effects have long been sought, but to date the only frequency dependent effects observed have been due to mass exchange between film and solution. The story starts with careful measurements of the q dependence of the propagation of capillary waves on 'solvent-free' BLM of GMO.[29] These experiments were carried out at 24.7°C — well above T_t. In this system the wave frequency and damping over a range of q (786–1602 cm^{-1}) both displayed small but significant deviations from the dispersion behaviour expected for constant membrane tension and viscosity. When the light scattering data were analysed to give membrane tension and viscosity, these quantities showed frequency variations which, within the precision of the data, were compatible with a single Maxwell viscoelastic fluid relaxation, having relaxation time ~ 37 μs. The equilibrium membrane tension deduced was compatible with expected values.

A similar study[15] of the frequency (i.e. q) dependence of the surface properties of a fully compressed monolayer of GMO at (20°C, again well above T_t) yielded similar conclusions: the transverse shear modulus γ again showed viscoelastic relaxation (Figure 5). As noted above, both experiment and data analysis were rather different for monolayers than for BLM so that the basic

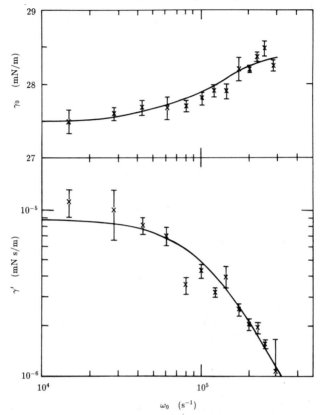

Figure 5: The frequency dependence of the surface tension and monolayer viscosity for a fully compressed film of GMO. The lines represent the best-fit Maxwell fluid behaviour, of relaxation time ~ 9 μs.

agreement of the phenomena for the two cases is quite reassuring: the effects seem to be real. The faster relaxation in the monolayer case $(\tau = 9$ μs) is not unexpected in view of the steric effects of the apposing monolayers in the BLM case. The fitted value of the equilibrium modulus again agreed exactly with the measured equilibrium surface tension, demonstrating conclusively that slower relaxation processes than those observed do not affect the transverse shear modulus.

These experiments were extended to more expanded monolayer states. As the area per molecule (A) increased from the fully compressed state, the difference between the light scattering and equilibrium values of the tension fell (tending to zero more rapidly at low q). This suggested that either the strength of the relaxation (G) or the relaxation time τ must fall as A decreases (c.f. Eqn. 4). However a decrease in G would result in a reduction in the viscos-

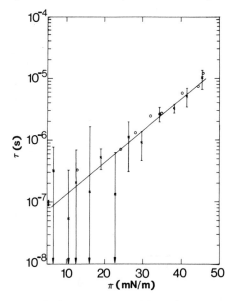

Figure 6: The variation of relaxation time with surface pressure for a monolayer of GMO. Data for $q = 1161$ cm^{-1} ($\times$) and for $q = 945.1$ cm^{-1} ($\circ$) are shown.

ity γ' on monolayer expansion (c.f. Eqn. 5). Experimentally, γ' was found to increase continuously on monolayer expansion until a rather expanded state ($A \sim 300$ Å^2) was reached. As A became still larger γ' fell, reaching zero for the clean subphase. It was thus inferred that the relaxation speeded up as the monolayer was expanded. Data observed for several different q values were consistent with the variation with π shown in Figure 6. Now π varied linearly with surface concentration of GMO over the range of interest, so that the variation of τ with π shown suggests that the relaxation speeds up exponentially as the molecular area is changed: τ changes by a factor of e for a 10% change in $1/A$. Such an exponential change would suggest that a single process is involved at all molecular areas. In the expanded phase ($\pi \sim 1$–2 mN/m, $A \sim 70$ Å^2) the relaxation must be very fast ($\tau < 10^{-8}$ s), so that all of the frequency dependence of γ occurs far above the present experimental range of ω. All of the data available to date are consistent with the hypothesis that the modulus γ involves only a single exponential relaxation of stress.[15]

The speeding up of the relaxation on monolayer expansion requires a concomitant rise (Figure 7) in its strength, G, to explain the observed increase in γ'. In fact it appears that G must become very large as τ drops: at $A \sim 300$ Å^2, $G \gtrsim 500$ mN/m. Thus in this regime of molecular area the effective tension of the monolayer will be very large indeed for fluctuations for which $\omega_0 > \tau^{-1}$. This will tend to reduce the amplitudes of thermally excited capillary waves of

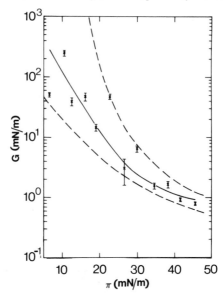

Figure 7: The variation of G with surface pressure for the same data as Figure 6. The lines indicate a plausible trend, and upper and lower limits.

such high frequencies (i.e. short wavelengths), with observable consequences for membrane processes in which such fluctuations are implicated.

The temperature dependence of the viscoelastic relaxation would be interesting, particularly around the membrane transition. Unfortunately, comprehensive studies of fluctuations over a range of q have not yet been performed at different temperatures. However, in a very detailed light scattering experiment on 'solvent-free' BLM the temperature dependence of γ was measured (for a single q value).[24] Under certain assumptions — that there is only one relaxation process, and that the equilibrium BLM tension as a function of T can be estimated from independent data — this suffices to determine the temperature variation of τ (Figure 8).[30] Remarkably, the relaxation of the transverse shear stress *speeds up* in the neighbourhood of the transition, in contrast with several previous studies of bilayer relaxation.[31] However, the light scattering approach addresses well-defined modes of motion of the membrane, unlike most other techniques used to investigate membrane relaxation, so that the conflict may be more apparent than real.

Thus far we have considered viscoelastic relaxation as far as it concerns the transverse shear modulus, γ. In the monolayer case, light scattering experiments are also sensitive to the dilational modulus, ε. This also displays relaxation,[15] but whereas for γ the relaxation is most noticeable for rather compressed monolayers, the frequency dependence of ε is found in the more expanded states

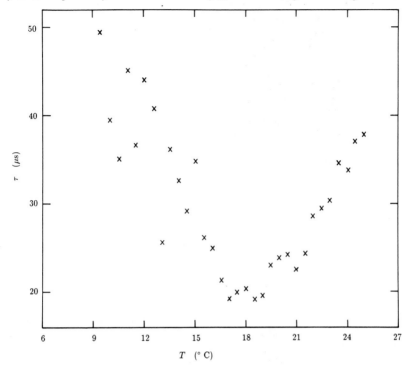

Figure 8: The variation of relaxation time with temperature for BLM of GMO. The transition occurs at 16.6°C. The three points at lowest T are unreliable.

$(60 < A < 700 \text{ Å}^2)$. In this regime the light scattering values of ε_0 exceed classical estimates by about 10 mN/m, essentially independent of q, while ε' is rather small, tending to fall with q (Figure 9). This sugggests a relaxation time scale slower than the capillary wave fluctuations. The ε' data are, within the large errors, compatible with a Maxwell model having relaxation time $\sim 100\ \mu s$, much slower than the relaxation of γ in this monolayer state. Other variations *cannot* be ruled out in this case. Capillary waves are rather insensitive to ε, particularly the viscosity ε': the values of ε' deduced from the light scattering data are thus very imprecise, so the relaxation is not very well defined. More precise experiments are needed. For a more compressed monolayer ε' is about an order of magnitude larger, suggesting that in that state the corresponding relaxation is rather faster than the inverse of the capillary wave frequency.

5 DISCUSSION

The results summarised above have shown how light scattering has opened an entirely new field of high frequency relaxation in model membranes. The mem-

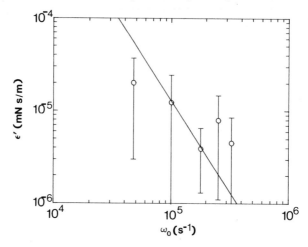

Figure 9: The frequency dependence of ε' (averaged over molecular area) for expanded GMO monolayers ($A > 60$ Å^2). The line shows the ω_0^{-2} dependence expected for a Maxwell model of $\tau \sim 100$ μs.

brane properties probed are clearly not thermodynamic, equilibrium values, but this apparent drawback in fact provides a wealth of new information. The time scales accessible — $\sim$ ms to $< \mu$s — cover a wide range, appropriate to a variety of molecular motions which must be significant for membrane function.

At the present time, because of the lack of a direct relation between the hydrodynamic modes of the membrane and microscopic motions, it is not possible to link the observed relaxation phenomena to molecular mechanisms directly. However some reasonably firm conclusions can be drawn for both transverse shear and in-plane dilation. In summary, the former case involves a relaxation, rather slower in BLM than in monolayers, which speeds up by more than three orders of magnitude as A is increased from the collapse area (28.5 Å^2) to $\sim$70 Å^2 and which, for BLM, displays a minimum at the transition temperature. Computer simulations show capillary wave-like modes associated with transverse motions of lipid molecules, involving chain-melting.[31] A time scale $\sim \mu$s has been suggested for chain-melting, from both experiments[32] and simulations,[33] in reasonable accord with the present observations for BLM or fully compressed monolayers. Increasing the molecular area will increase the conformational freedom of the lipid molecules, permitting easier melting and hence more rapid relaxation of transverse shear stress.

'Bobbing up and down' molecular motions such as are involved in the capillary waves have been associated with membrane permeability,[1,32] which is known to exhibit a transitional maximum.[34] This maximum has been explained by differences in ease of permeation through 'solid', 'liquid' and 'interfacial' (between solid and liquid domains) regions of the bilayer.[35] Computations to date have not provided a rationale for such differences, but if they are due to different rates of transverse motions of the molecules, the observed minimum in τ may rather naturally be explained. This mechanism might also explain the fluctuations evident in Figure 8 below T_t: below the transition the 'solid' domains may become relatively large, so that the amount of 'interfacial' lipid within the illuminated area will tend to fluctuate, leading to changes in the rate at which the local transverse shear stress can relax.

The relaxation involved in the case of the dilational modulus occurs at large A, extending to areas where the surface pressure is negligible, and the molecules are, on average, widely separated. It appears that the dilational motions associated with the capillary waves, which are of relatively high frequency (compared to slower perturbations, such as the quasi-static compression involved in the classical determination of ε), perceive the monolayer as *relatively* incompressible. This is compatible with the formation of some local supra-molecular structure in very expanded monolayers, which can easily be compressed slowly, but which resists rapid dilational stress. Such structures, including surface 'foams'[36] and 'super-lattices',[37] have been postulated: the present data appear able to accomodate either model.

The light scattering technique would probably best be used to study changes in membrane properties, such as at transitions or following chemical modification of the membrane. An obvious and interesting example would be the effects of membrane modifiers on 'fluidity', or rather the relaxation of the well-defined viscoelastic properties accessible to light scattering. Unfortunately, early demonstrations[38,27] of the cholesterol induced increase of membrane viscosity (γ') were insufficiently precise to permit analysis of any relaxation involved. Correlation of molecular relaxation with membrane processes would substantially increase the information content of the results.

ACKNOWLEDGEMENTS

This work has been supported by the Science and Engineering Research Council.

REFERENCES

1. R.N. Robertson, 'The Lively Membranes', Cambridge University Press, Cambridge, 1983, pp. 59–79.
2. L. Kramer, J. Chem. Phys., 1971, 55, 2097.
3. F.C. Goodrich, J. Phys. Chem., 1962, 66, 1858.
4. D. Sornette and N. Ostrowsky, J. Chem. Phys., 1986, 84, 4062.
5. A. Milon, J. Ricka, S.-T. Sun, T. Tanaka, Y. Nakatani and G. Durisson, Biochim. Biophys. Acta, 1984, 777, 331.
6. M.B. Schneider, J.T. Jenkins and W.W. Webb, J. Phys. France, 1984, 45, 1457.
7. S.B. Hladky and D.W.R. Gruen, Biophys. J., 1982, 38, 251.

8. M.A. Bouchiat and D. Langevin, J. Coll. Interface Sci., 1978, 63, 193.
9. F.C. Goodrich, Proc. R. Soc. London A, 1961, 260, 503.
10. F.C. Goodrich, Proc. R. Soc. London A, 1981, 374, 341.
11. M. Baus, J. Chem. Phys., 1982, 76, 2003.
12. J.C. Earnshaw and R.C. McGivern, J. Phys. D, 1987, 20, 82.
13. D. Langevin, J.Meunier and D. Chatenay, 'Surfactants in Solution', K.L. Mittal and B. Lindman (eds.), Plenum, New York, 1984, Vol. 3, p.1991.
14. J.C. Earnshaw, Thin Solid Films, 1983, 99, 189.
15. J.C. Earnshaw, R.C. McGivern and P.J. Winch, J. Phys. France, 1988, 49, 1271.
16. J.C. Earnshaw, 'Inverse Problems in Optics', E.R Pike (ed.), 1987 Proc. SPIE 808, p. 158.
17. J.D. Ferry, 'Viscoelastic Properties of Polymers', John Wiley, New York, 1980.
18. C.P. Fan, J. Coll. Interface Sci., 1973, 44, 369.
19. F. Brochard, P.G. de Gennes and P. Pfeuty, J. Phys. France, 1976, 37, 1099.
20. R.B. Tishler and F.D. Carlson, Biophys. J., 1987, 51, 993.
21. J. Meunier, J. Phys. France, 1987, 48, 1819.
22. G.E. Crawford and J.C. Earnshaw, J. Phys. D, 1985, 18, 1029.
23. P.J. Winch and J.C. Earnshaw, J. Phys. E, 1988, 21, 287.
24. G.E. Crawford and J.C. Earnshaw, Biophys. J., 1986, 49, 869.
25. J.C. Earnshaw, 'The Application of Laser Light Scattering to the Study of Biological Motion', J.C. Earnshaw and M.W. Steer (eds.), Plenum, New York, 1983, p. 275.
26. J.C. Earnshaw, 'Laser Scattering Spectroscopy of Biological Objects', J. Štěpánek, P. Anzenbacher and B. Sedláček (eds.), Elsevier, Amsterdam, 1987, p. 469.
27. G.E. Crawford and J.C. Earnshaw, Eur. Biophys. J., 1984, 11, 25.
28. J.C. Earnshaw and R.C. McGivern, to be published.
29. G.E. Crawford and J.C. Earnshaw, Biophys. J., 1987, 52, 87.
30. J.C. Earnshaw and G.E. Crawford, to be published.
31. A. Genz and J.F. Holzwarth, Coll. Polymer Sci., 1985, 263, 484.
32. B. Owenson and L.R. Pratt, J. Phys. Chem., 1984, 88, 6048.
33. T. Lookman, D.A. Pink, E.W. Grundke, M.J. Zuckermann and F. deVerteuil, Biochemistry, 1982, 21, 5593.
34. D. Papahadjopoulos, K. Jacobsen, S. Nir and T. Isac, Biochim. Biophys. Acta, 1973, 311, 330.
35. L. Cruzeiro-Hansson and O.G. Mouritsen, preprint, 1988.
36. B. Moore, C.M. Knobler, D. Broseta and F. Rondolez, J. Chem. Soc., Faraday Trans. 2, 1986, 82, 1753.
37. D. Andelman, F. Brochard, P.G. de Gennes and J.-F. Joanny, C. R. Acad. Sci. Paris, 1985, 301, 675.
38. J.F. Crilly and J.C. Earnshaw, Biophys. J., 1983, 41, 211.

Index

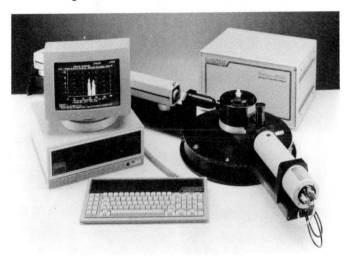